The Open University

S357 S

Space, Time and Cosmology

BLOCK 4

Cosmology and the early Universe

Unit 13 The big bang

Unit 14 General relativity and cosmology

Unit 15 The evolution of the Universe

Unit 16 Consolidation and revision II

S357 Course Team

Course Team Chair: Raymond Mackintosh

Raymond Mackintosh	Author	Michael Watson	Course Manager
Russell Stannard	Author	David Tillotson	Course Editor
Bob Zimmer	IET	Peter Twomey	Course Editor
Tony Evans	Author, Assessor	Ian Thomas	Producer, BBC/OUPC
Leon Firth	Author	Liz Sugden	Assistant, BBC/OUPC
Bernard Schutz	Author	Tony Jolly	BBC/OUPC
John Charap	External Assessor	Steve Best	Graphic Artist
Gillian Stansfield	Assessor	Sarah Crompton	Graphic Design
Tom Smith	Reader	Hannah Brunt	Graphic Design
Alan Cooper	Reader	Alison Cadle	Manager OU TeX system

S357 is a revision of a course, S354, first presented in 1979. We should particularly like to acknowledge major contributions to the original course team members: John Bolton (OU), David Broadhurst (OU), Paul Clark (OU), Alan Cooper (OU), Tom Smith (OU), Russell Stannard (OU) Andrew Crilly (BBC), Al Saperstein, George Abell, Julian Schwinger.

2004 Revision Team

David Broadhurst (Chair)
Robert Lambourne (Author)
Raymond Mackintosh (Reader)
Michael Watkins (Course Manager)

List of Units

Block 1   Newtonian ideas of space and time
Unit 1   Space and time — the setting for the motion of a particle
Unit 2   Newtonian mechanics
Unit 3   Symmetrical and conservation laws

Block 2   Electromagnetism and Einstein's special theory of relativity
Unit 4   Electromagnetic fields
Unit 5   The need for special relativity
Unit 6   Some consequences of special relativity
Unit 7   Spacetime, momentum and energy
Unit 8   Consolidation and revision I

Block 3   Gravitation, Einstein's general theory of relativity and black holes
Unit 9   First steps to a theory of gravitation
Unit 10 & 11   A metric theory of gravity and the field equations of general relativity
Unit 12   Black holes and other consequences of general relativity

Block 4   Cosmology and the early Universe
Unit 13   The big bang
Unit 14   General relativity and cosmology
Unit 15   The evolution of the Universe
Unit 16   Consolidation and revision II

The Open University, Walton Hall, Milton Keynes, MK7 6AA

First published 1997.

Edited, designed and typeset by the Open University using the Open University TeX System.

Printed in the United Kingdom by Martins the Printers, Berwick upon Tweed.

ISBN 0 7492 6981 2

This text forms part of an Open University Third Level Course. If you would like a copy of *Studying with The Open University*, please write to the Course Enquiries Data Service, PO Box 625, Dane Road, Milton Keynes, MK1 1TY. If you have not already enrolled on the Course and would like to buy this or other Open University material, please write to Open University Educational Enterprises Ltd, 12 Cofferidge Close, Stony Stratford, Milton Keynes, MK11 1BY, United Kingdom.

2.1

S357b4i2.1   *Title page photo of Friedmann is from a memorial article in the 1963 Soviet journal Usp. Fiz. Nauk.*

Unit 13 The big bang

Prepared by the Course Team

Contents

Aim

In this Unit, we present the three main lines of experimental evidence pointing to the big bang origin of the Universe: (i) the recession of the galaxies; (ii) the microwave remnant of the early fireball; and (iii) the comparison between the calculated primordial nuclear abundances and the present-day composition of matter in the Universe.

Objectives

When you have finished studying this Unit, you should be able to:

1 Describe the characteristics of light emitted by stars, and hence the information of cosmological interest that can be deduced from it.

2 Distinguish between true and false statements relevant to the distribution and motion of stars within galaxies, and of galaxies within clusters and superclusters.

3 Outline the methods used for estimating the distances to stars and to galaxies.

4 Explain and apply Hubble's law.

5 Distinguish between various sources of redshift and estimate their relative importance in a given situation.

6 Describe the cosmic microwave radiation in terms of its origin, its thermal nature, and how its temperature has changed with time.

7 Explain how observations of thermal radiation are affected by the motion of the observer with respect to it.

8 Outline the evidence concerning the isotropy of the Universe.

9 Describe the most important basic nuclear reactions taking place in the early stages of the big bang, and how the final mix of elements produced provides evidence today for the big bang.

1   Introduction

Band 1 of AC4 introduces this Unit.

General relativity has a very different conceptual basis from that of Newtonian mechanics. Its success in accounting for the precession of Mercury's orbit, and the bending of light by massive objects like the Sun, gives us confidence that our picture of space and time should be Einstein's rather than Newton's. In this and the following Units, we turn our attention to the study of the large-scale structure of spacetime. We see how spacetime as a whole is curved by the gross distribution of mass and energy in the Universe. This distribution, together with the question of how the Universe has developed over time, is the subject of *cosmology*.

cosmology

Astronomy and cosmology are subjects that merge into one another with the single combined aim of understanding the structure and history of the Universe. The basis and motivation for the whole subject area comes ultimately from astronomers' observations. Since the days of Galileo, optical telescopes of ever greater size have been made. In the last 60 years, it has been possible to study an increasing range of the electromagnetic spectrum as different types of telescope have become available. The first radio telescopes were made just before World War II. Infrared, ultraviolet, X-ray and gamma ray telescopes then followed, often operating from spacecraft above the Earth's atmosphere. These methods have become so complicated that we have attempted to outline only certain results; a far longer course would have been necessary to do anything like justice to the delicacy and sophistication of the techniques involved.

What have we learned from these new techniques? Firstly, matter has been detected in a wide variety of forms — interstellar gas as well as the stars themselves, for instance. Secondly, it has become possible to perform detailed studies of the radiation from the very ancient Universe. These developments have given cosmologists a more comprehensive list of the forms of mass and energy that govern the spacetime of the Universe. Next, matter has been detected at much greater distances — the more distant view providing cosmologists with more telling tests of their models. Then there has been the increasingly detailed information on spectra, with its evidence on the compositions of the emitting bodies. This information is of particular concern to the astrophysicist, who tries to understand the evolution of stars and other types of matter. Such evolution is governed by stellar dynamics and the processes of nuclear physics. The latter will be touched on later, though with reference more to nuclear reactions taking place in the very early Universe than to those occurring in the stars today.

We hope in this Block to build a bridge between the two extremes — the raw data obtained by the astronomer, and the metric parameters derived by the cosmologist.

In this first Unit, we examine the reasons for believing that the Universe had its origins in the big bang. We shall find that there are three independent pieces of evidence all pointing to the same conclusion. Sections 2 to 4 introduce you to the first of these.

4

2 Radiation from the galaxies

galaxy

Stars occur in great collections called *galaxies*. The distribution and motion of galaxies provides us with the first important experimental information on which we shall build our understanding of the type of universe we inhabit. So, what do we know about galaxies?

All the stars that can be distinguished by the naked eye — a few thousand in number — belong to one galaxy: our own Milky Way Galaxy. Sometimes it is just written Galaxy, with a capital G, to distinguish it from all the billions of other galaxies in the observable Universe.

Our Galaxy has the overall shape of a flattened spiral. As shown in Figure 1, we are halfway inside the Galaxy so the spiral form is not obvious to us as we look at the sky. In fact it has only been revealed by detailed radio mapping. This is because there are large sections of the Galaxy that are obscured from optical observations by the intervening *dust* and gas which scatter and absorb visible light. (However, they are essentially transparent to radio waves.) If there were an unobscured view through telescopes, we would be able to see to the far side of the Galaxy, and in doing so we would record about 10^{11} stars within it.

galactic dust

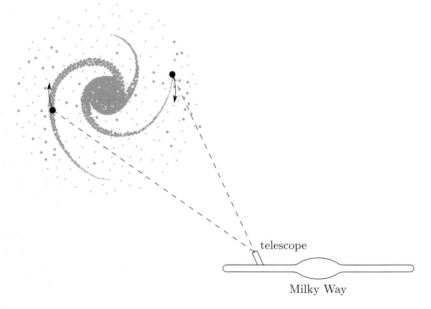

Figure 1   The Solar System is about halfway out from the centre of the Milky Way galaxy. We have a relatively clear view outwards from the disc to other galaxies.

Fortunately for astronomers, the stars, though vast in number, display a degree of uniformity which makes their classification and study possible. Their masses are broadly similar — most are contained within a couple of orders of magnitude, with our Sun having a typical average mass. They do differ considerably in size and therefore in density (since their masses are similar), but these differences are thought principally to relate to different *stages* in the life of a star rather than to different *types* of star. These various stages correspond to different types of nuclei taking part in the nuclear reactions that provide the energy output. It is one of the features of astrophysics that one is able to take nuclear data obtained in the laboratory and use them to understand the various stages of evolution seen in different stars. The main process in young stars is a sequence of reactions leading to the fusion of hydrogen nuclei to make helium nuclei. Later, helium fuses to form carbon, then carbon undergoes reactions which build up heavier nuclei. This can continue up to iron. The different reactions take place at different temperatures and pressures, so this progression of reactions governs the sequence in which a star changes its size and appearance.

Stellar evolution is too slow a process for us to see any particular star undergo change in one human lifetime, apart from a few exceptionally young stars, and some very old stars undergoing gravitational collapse leading to supernova explosions. But by observing different stars at their respective different stages of development, it is possible to piece together the whole of a typical stellar life cycle.

evolution of galaxies

The general idea of an evolutionary sequence — one that can be reconstructed on a computer — is of concern in this Unit for the following reason: when astronomers look at a very distant galaxy, they are receiving radiation that left its source long ago. The galaxy will, therefore, seem *younger* than it actually is now. The travel time for the light may have been hundreds or thousands of millions of years. To interpret the observations, one needs to know how the power output of the galaxy evolves with time.

The light from a typical galaxy derives mostly from the stars it contains, with only a small amount from interstellar matter. So, to understand the power output of a galaxy one has to add together the light from about 10^{11} stars at various stages of development. We cannot assume that the stars of the distant galaxy will be emitting the same amount of light as those of our own Galaxy today; the distant stars will be seen at an earlier stage in their development when perhaps their power output was different from what it actually is now. Indeed, there is an added complication. The evolution of a star depends critically on its *mass*. A very massive star will shine much more brightly than a less massive one, but over a much shorter time period. It could be that when we look at a young galaxy, we see many more massive stars living out their brief active lives than we observe today in our own Galaxy. So the mass distribution of active stars in our Galaxy might not be representative of what we see happening in the distant younger galaxy. It is difficult to know how to compensate for this. The mass distribution of the stars in a galaxy depends on the way the galaxy was formed, and unfortunately, the formation and growth of galaxies remains an unsolved, or at least poorly understood, problem.

In summary, we know little about the way the power output of a galaxy changes with time, and this represents a severe limitation on the cosmologist's use and interpretation of astronomical data on power output. The frustration this causes will become apparent later.

spectrum

The light from a galaxy can tell us more than just its overall power output. Additional information comes from the analysis of its *spectrum*. Let us assume that light comes mainly from the stars and that we can ignore interstellar matter. The majority of galaxies are too far away for it to be possible to resolve individual stars, and therefore the best one can do is to take the spectrum of the galaxy as a whole: 10^{11} stars summed together. How can we relate this to a stellar spectrum?

The light from a star comes from the hot layers of gas near its surface. This light filters out through the dilute layers above. As it does so, characteristic patterns of absorption lines are imprinted on the spectrum.

6

absorption spectrum

emission spectrum

Figure 2 shows part of the *absorption spectrum* of the Sun, compared to an *emission spectrum* produced in the laboratory.

Objective 1

SAQ 1   A beam of white light is passed through a bulb containing calcium vapour. Explain, in broad terms, with the aid of diagrams, the nature of the light transmitted, (a) in the direction of the beam, and (b) at an angle to it.

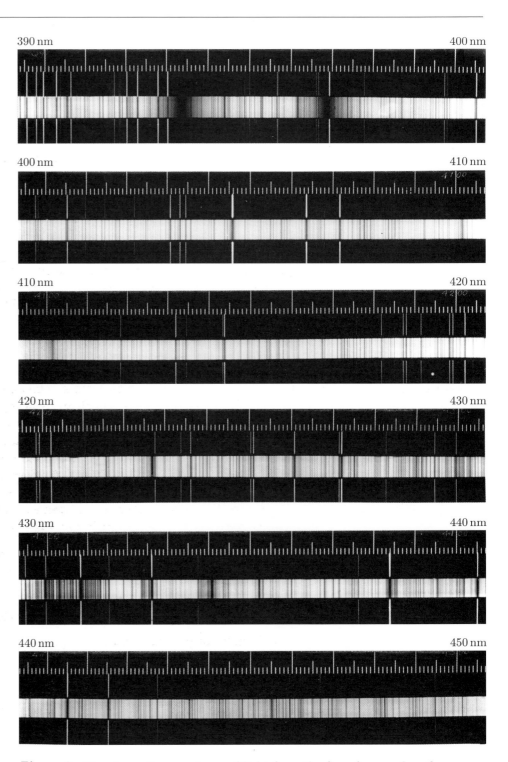

Figure 2   The absorption spectrum of light from the Sun, for wavelengths between 390 nm and 450 nm. It is complicated because of the many elements involved. The bright lines above and below the absorption spectrum belong to the emission spectrum of a laboratory reference source.

The spectral lines in the light from a galaxy enable the astronomer to identify the elements emitting the light. Any of the ninety or so stable elements may be present, but the lighter elements, especially hydrogen, are usually the more abundant. Because some elements are common to most stars, the absorption lines of these particular elements will be visible in the light of the galaxy as a whole. Also, the absorption lines of magnesium, sodium, calcium and iron are often easy to distinguish, even though other elements, such as hydrogen and helium, are more abundant than these.

Doppler shift

Usually the absorption lines are sharp and identifiable, despite several effects which can broaden them. One of these effects is a *Doppler shift* caused by the rotation of the source. For example, the spectral lines of calcium atoms moving towards us would have their apparent frequencies systematically increased (blueshifted) according to Equation 5 of Unit 6, and those of atoms moving away from us would be correspondingly redshifted. This Doppler shift causes the width of a given spectral line to *broaden* if we are viewing the rotating galaxy along the plane of its disc so

Doppler broadening

that the light-absorbing atoms belonging to stars on one side are moving away from us, while those on the opposite side are moving towards us (refer back to Figure 1). In addition, the star itself might be rotating so that different parts of the star would have different components of velocity towards or away from us, thereby increasing the broadening. Random motions of stars can also cause line broadening in a similar way.

The shifts and broadenings of the lines are not usually bad enough to mask the spectrum. Therefore, although a galaxy is a very complicated light source, its light is not just a meaningless jumble of fuzzy lines. There are some features, such as the calcium lines, which stand out sharply.

Objectives 1 and 5   **SAQ 2**   (a) It was shown in Unit 9 that when radiation rises a height H near the Earth's surface, there is a shift in frequency towards the red end of the spectrum given by $\Delta f/f = -gH/c^2$, where g is the acceleration due to gravity. When radiation starting from the surface of a body of mass M and radius R escapes to a large distance, it suffers a fractional frequency shift $\Delta f/f = -GM/Rc^2$, where G is the gravitational constant. Outline the steps you would need to take to establish the connection between these two statements *without* necessarily giving the derivation. (*Hint*: Bear in mind that $g = GM/R^2$.)

(b) Imagine a star with a mass $M = 2 \times 10^{30}\,\text{kg}$, a radius $R = 7 \times 10^8\,\text{m}$ and a period of rotation $T = 2 \times 10^6\,\text{s}$ (similar to the Sun). Suppose also that the typical speed of turbulent motion in the atmosphere of the star is $6000\,\text{m s}^{-1}$. How is the frequency of the hydrogen line whose wavelength in the laboratory is 656 nm, affected by:

(i) gravitational redshift;

(ii) Doppler shift due to rotation;

(iii) Doppler shift due to turbulence?

(The values of G and c are given on the back cover of the Block, and the formula for the Doppler shift of light is given by Equation 5 of Unit 6.)

In many galaxies, spectral lines have been observed whose *relative* positions correspond exactly with those of a known element (so that the spectrum is confidently identified), but whose *absolute* frequencies are all noticeably shifted. This can have nothing to do with the random motions of the stars within the galaxy, or the rotations of either the individual stars or of the galaxy as a whole; all these motions (being as often towards us as away from us) would merely broaden the line. A net shift of all the lines in a spectrum seems to suggest that the galaxy itself has a line-of-sight motion (i.e. a component of velocity towards or away from us). The shift is nearly always to longer wavelengths, i.e. towards the red end of the visible spectrum, so it is referred to as the 'redshift'. If the redshift is interpreted as a Doppler shift (an interpretation we shall reconsider later) due to the motion of the galaxies, then it follows that most of the galaxies are receding from us. In other words, *the Universe is expanding*.

The term 'redshift' has a precise definition: it is equal to $\Delta\lambda/\lambda$, where $\Delta\lambda$ is the shift in wavelength of a line whose emitted wavelength is λ. The value of $\Delta\lambda/\lambda$ is the same for all the lines in the spectrum of an object receding at a given speed, and is normally denoted by z:

$$z = \Delta\lambda/\lambda$$
$$= \frac{\lambda_0 - \lambda_1}{\lambda_1} \tag{1}$$

where λ_0 is the observed wavelength now, and λ_1 is the wavelength at emission. These redshifts in the spectra of galaxies are generally *far* larger than the stellar effects considered in SAQ 2, as can be judged from Figure 3.

Note that it is customary to reserve the suffix, 0, for the values of quantities as they are at the present time, t_0. Values at other times must carry some other suffix. In this way, we end up with the rather counterintuitive situation where λ_1 applies to an *earlier* time, t_1, than λ_0.

In summary then, when dealing with light from stars, there are three ways in which the frequency can be shifted:

(i) The Doppler shift due to motion — whether that arises through the rotation of the star, its bodily motion along the line of sight, or turbulence in its atmosphere. This type of shift, which can be red or blue, is accounted for by special relativity.

(ii) The gravitational redshift, arising from general relativity.

(iii) A new type of redshift due to the recession of the galaxies, which, as we shall see in Unit 14, is also a consequence of general relativity.

Objective 1   SAQ 3   Hydrogen has emission lines at 434, 486 and 656 nm. A galaxy is observed to have these three lines redshifted in wavelength by 2% (i.e. $z = \Delta\lambda/\lambda = 0.02$). What wavelengths will be observed? What is the frequency shift $\Delta f/f$?

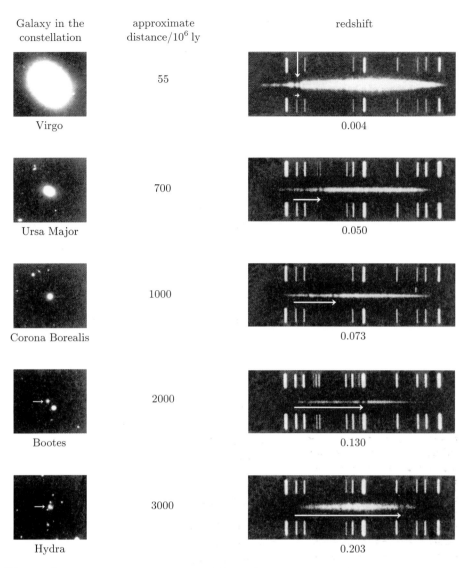

Figure 3 The absorption spectra of five different galaxies. Only the most strongly absorbed lines, common to the bulk of the stars, are visible, but this is enough to give a recognizable pattern.

As we shall see later in this Unit, the redshift in galactic light has provided one of the main clues to the nature of the large-scale structure of spacetime. But to discover how the pattern of the redshifts observed in different galaxies reveals this structure, we need further information about the distances of galaxies.

3 Distances of galaxies

3.1 First steps towards a distance scale

As you will see from Table 1, when it comes to astronomy and cosmology, one is called on to deal with a wide range of distances. (Note that a light-year (ly) is the distance light travels in one year, i.e. 9.46×10^{15} m. The distances are also quoted in a very commonly used astronomical unit of distance: the megaparsec, Mpc, where a parsec (pc) is 3.26 ly or 3.09×10^{16} m.)

TABLE 1

	Distance	
Earth–Moon	1.28 light-seconds	1.25×10^{-14} Mpc
Earth–Sun	8.3 light-minutes	4.8×10^{-12} Mpc
Nearest star	≈ 4 ly	1.23×10^{-6} Mpc
Diameter of our Galaxy	$\approx 10^5$ ly	3×10^{-2} Mpc
Distance to nearest galaxy	$\approx 2 \times 10^6$ ly	≈ 1 Mpc
Farthest galaxy seen	$\approx 4 \times 10^9$ ly	$\approx 1.2 \times 10^3$ Mpc

To measure the distance of a far-off galaxy clearly requires a series of steps. The first of these, the Earth–Sun distance, is based on our knowledge of the orbital period of the Earth about the Sun, and that of some other planet. One can then readily obtain an estimate of their *relative* distances from the Sun by using Kepler's third law. To convert this to an *absolute* measurement, one needs a determination of the actual distance between the Earth and the planet at some time. In practice, this fix is gained by measuring the distances to Mercury, Venus and Mars, using radar. These then allow one to compute the Earth–Sun distance. It is currently estimated, with obvious high precision, to be 149 597 870.66 km.

Knowing this, the diameter of the Earth's orbit can now be used as a baseline for measuring the distance to nearby stars, using the surveyor's triangulation method. One makes angular measurements on a star from opposite ends of an Earth-orbital diameter, i.e. at intervals of 6 months. At the time of writing (2004), data from the Hipparcos satellite surpass in precision and scope all previous measurements of nearby stellar distances.

But no matter how good this satellite-gathered data, there soon comes a stage where the angles become too small to measure accurately using the triangulation method. Some other method must be employed to extend the distance scale to the more distant stars.

luminosity   If a source has a known total light output, i.e. known *luminosity*, then it can be used as a standard source. The distance of the source can be found from the *received* light power. The further away the star is, the fainter it appears. The relationship follows an inverse square law (allowing for various corrections, such as that due to the absorption of light by interstellar dust). The light power received per unit area of detector is **flux density**   called the *flux density*. It follows, summing over the area of a sphere centred on the star and of radius, r, equal to the distance to the star, that in the absence of absorption:

$$\text{luminosity} = 4\pi r^2 \times \text{flux density}. \tag{2}$$

Unfortunately stars differ widely in their luminosities, so if we simply look up at the sky and pick out a faint star, it may be either an intrinsically dim star, or a particularly distant star — or partly dim and distant.

This ambiguity can be partly overcome by recognizing that, while it is impossible to directly measure the luminosity of a particular source, one *can* estimate the temperature of the star. This is done from measurements on the overall shape of its emitted spectrum — stars emitting predominantly in the red part of the spectrum being cooler than those that emit more in the blue region. Thus the spectrum, and hence temperature, can be determined with a fair degree of confidence. Now, it turns out that when observations are made on a compact cluster of stars (the cluster being small enough for all its stars to be considered equidistant from us), a plot of flux density versus temperature generally gives the *same* distribution no matter which cluster is chosen — apart from an overall constant factor depending on the distance to the cluster. Thus, the most likely luminosity of a star is related to its temperature. But in order to use the temperatures of stars to obtain luminosities, and hence distances, from Equation 2, we need to progress beyond a plot of *flux density* versus temperature to one of luminosity versus temperature.

In order to achieve this, studies have been made of a group of about 100 stars called the 'Hyades star cluster'. This group is close enough to us for its distance to be determined by the Hipparcos satellite. Knowing this distance and the flux density of the stars, the luminosities could be established, and the plot of luminosity versus temperature calibrated. By studying the flux densities and colours of stars more distant than the Hyades star cluster, we can use this calibrated plot to infer their distances from us.

Cepheid variable

This method applies to stars confined to our own Galaxy. As for stars in other galaxies, most of these cannot be distinguished separately. So, the next step is to try to find a particularly bright type of star — one that *can* be recognized not only in our Galaxy but also in neighbouring galaxies. If this were possible, it would give us a method of extending the distance scale out to other galaxies. Fortunately there *does* exist a type of star that can be recognized at least in neighbouring galaxies — a star known as a *Cepheid variable*. The important characteristic of a Cepheid is that its light output varies in a regular fashion, with a period which is directly related to its mean luminosity (see Figure 4a). The variation is due to cyclical changes in diameter. The relationship between mean luminosity and period can be calibrated by studying Cepheids that are sufficiently close for their distances to have been measured by methods mentioned earlier. So whenever such stars can be distinguished from the other types of variable star in the more distant galaxies, their luminosities can be deduced (see Figure 4b). Comparing luminosity with the measured flux density then enables the distance of the star to be calculated. Cepheids have been used extensively to measure the distances to nearby galaxies — those belonging to a cluster of galaxies known as the Local Group (to be described later). Following the advent of the Hubble Space Telescope with its superior resolution of individual stars, it became possible in 1996 to extend Cepheid-based measurements as far as a galaxy known as M100, in the Virgo cluster, yielding a distance of 5.6×10^7 ly.

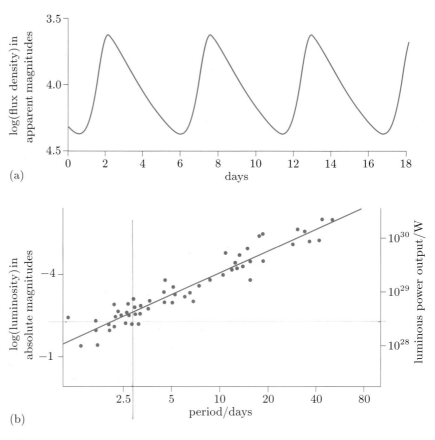

Figure 4   (a) An 18-day section of the light curve of the typical Cepheid variable, Delta Cephei, which has a 5.37-day oscillation. (*Apparent magnitude* is a term defined in the Glossary.)

(b) The observed relationship between the period and luminosity (in watts) of Cepheid variables ($1\,\text{W} = 1\,\text{J}\,\text{s}^{-1}$). Notice that both scales are logarithmic, so the straight line implies that period $\propto$ (luminosity)^{m}, where $1/m$ is the slope of the line. (*Absolute magnitude* is a term defined in the Glossary.)

apparent magnitude

absolute magnitude

Type Ia supernova

Another type of star that can be recognized in other galaxies is a *Type Ia supernova*. What happens is this: when a medium-sized star, such as the Sun, approaches the end of its active life, it shrinks down to a small star called a 'white dwarf'. If the white dwarf happens to belong to a binary system of two stars, it can, from time to time, capture material from the atmosphere of its companion, so increasing its own mass. But this is a process that cannot continue indefinitely. The maximum mass for a white dwarf is 1.4 solar masses; anything above that limit and its inner forces cannot resist the inward pull of gravity, and the star has to collapse down to the next stable form (called a 'neutron star'). Thus, the white dwarf in the binary system can capture material only up to this limiting mass. Once it exceeds this limit, the collapse occurs with the excess energy emitted as an explosion — the Type Ia supernova. Because white dwarfs are always of the same limiting mass when this happens, the explosions are similar, yielding essentially the same spectrum and variation of light intensity over time. (It is these characteristics that allow the Type Ia supernovae to be distinguished from other types.) Why they are important in the present context is that, to within $\pm 30\%$, they have the same luminosity. Taking into account that the variation in luminosity with time does show a difference in characteristic decay times, and these are correlated to somewhat different values of peak luminosity, allowance can be made for this, and the uncertainty spread in peak luminosity reduced to $\pm 20\%$. This in turn leads to relative distances measurable to $\pm 10\%$.

Type Ia supernovae provide us with standard lamps (or standard *explosions*!), that, at their peak, are 100 000 times brighter than a Cepheid, and are visible hundreds of times further away. Having established their luminosity by measuring the few that occur in galaxies for which the distance is already known from measurements on Cepheids, they can be used to extend the distance scale to very great distances.

This however still does not go far enough. Since normal stars cannot be resolved in the farther galaxies, the additional methods of estimating distance will have to be based on the properties of galaxies *as a whole*. We therefore interrupt the distance story to describe some properties of whole galaxies, showing first why they are important to cosmology, and then how they have led to new ways of measuring distance.

Objectives 1 and 3   SAQ 4   Suppose an astronomer using a telescope of 2 m diameter has a detector whose limit of sensitivity is 3.2×10^{-17} W. Use Figure 4 to deduce, for this instrument, the period of the faintest Cepheid variable that can be observed at a distance of 2×10^6 ly.

(*Remember*: The value of a light-year in SI units is given on the back cover.)

3.2   Some general properties of galaxies

cluster of galaxies

Firstly, we note that galaxies tend to occur in clusters rather than singly. The mutual gravitational attraction of galaxies naturally tends to hold them on paths that remain close to each other. Typically a cluster contains tens or hundreds of galaxies. There are, however, large clusters with thousands of galaxies, and there are some solitary galaxies. Our own Galaxy is a member of a smallish cluster of about 36 galaxies called the Local Group (see Figure 5). A typical cluster of moderate size is shown in Figure 6.

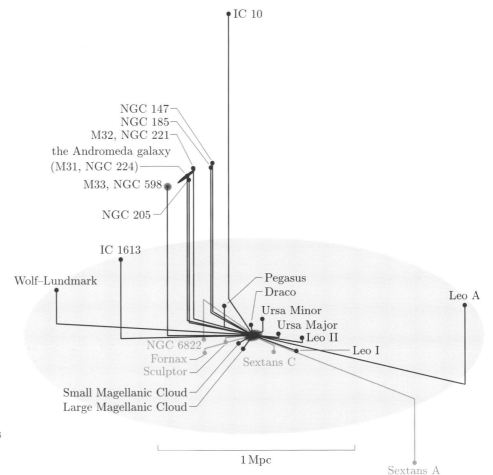

Figure 5 Our cluster of galaxies, called the Local Group. M31, the Andromeda galaxy, can be seen with binoculars. The Large Magellanic Cloud is one of our nearest neighbours, and is visible to the naked eye from the Southern Hemisphere.

Figure 6 A cluster of galaxies whose position in the sky is behind the constellation of stars in our Galaxy which we call Hercules. This cluster is not visible to the naked eye — not because it is too small but because it is too faint.

How are the galaxy clusters distributed? Are they close enough to affect each other? If they are, then the motion of a given cluster would depend mostly on the distortion of spacetime caused by its nearest neighbours. Because the distances between clusters vary, this would correspond to a large spread in relative speeds. At the other extreme, if the clusters of galaxies were very far apart, then the attraction of neighbouring clusters would be negligible, and the motion would be dominated by the overall spacetime curvature due to the matter of the *whole* Universe — a much simpler situation.

As we shall see in Unit 15, galaxy clusters are loosely associated in superclusters. However, they are far enough apart for one to regard clusters as essentially independent (see Figure 7). The fact that clusters are, in effect, independent of one another is of central importance to cosmology. It means that a cluster of galaxies can be taken as the basic 'particle' of cosmological dynamics, and the motion of individual galaxies within a cluster can be ignored on the grand scale of cosmology. Thus, we conclude:

Galaxy clusters are the basic test particles of cosmology, their motion following geodesic paths through spacetime.

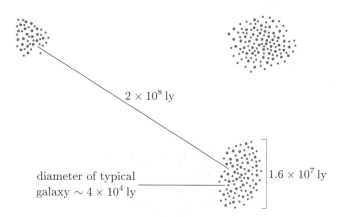

Figure 7   A very schematic view of three clusters, showing typical diameters of galaxies (single dots) and clusters (groups of dots), and typical distances between clusters.

In other words, a galaxy cluster plays a similar role on the grand scale to that of a planet (e.g. Mercury) mapping out the local region of spacetime in the Solar System, or Newton's falling apple doing the same thing closer to the Earth's surface.

The next important property of galaxies is that there is a statistical predictability about their masses and luminosities. Figure 8 shows the proportion of galaxies having a given luminosity. It varies over quite a wide range, the distribution falling off steeply on the high-luminosity side. (Note that the luminosity is expressed in terms of (absolute) 'magnitudes', this being a parameter such that the *smaller or more negative* the value of the magnitude, the more luminous the object. This rather odd choice arises for historical reasons.) There are large numbers of the least luminous galaxies (shown by the rising curve on the left-hand side), but in practice these tend to be invisible in the more distant clusters which are of greater interest to cosmologists. No doubt it would have been easier for cosmology if galaxies had been more similar. However, because galaxies occur in clusters, one can at least use statistical methods. For instance, the distribution of luminosities is more or less the same for all clusters, resulting in the average luminosity of galaxies in a cluster being fairly standard even though the individual galaxies vary greatly. This statement assumes, of course, that we are dealing with galaxy clusters of the same age. As we have already pointed out, this might not be the case when observing galaxies at a great distance, and hence as they were some time in the distant past.

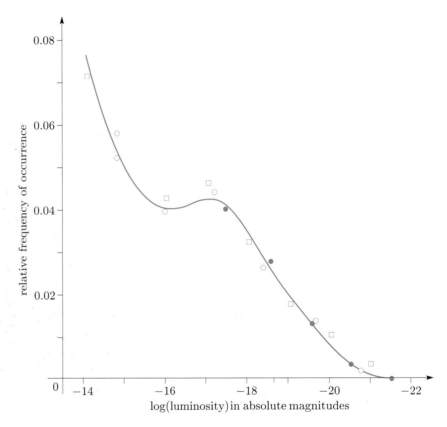

Figure 8   Number of galaxies per magnitude class per cubic megaparsec, as a function of absolute magnitude. Different symbols represent the results of different observers.

3.3   Extending the distance scale

Having reviewed some of the properties of galaxies, we are now in a position to return to the question of how we are to develop further our methods of measuring distance.

The various steps taken in determining larger distances from known smaller ones are often called 'rungs in the distance ladder'. The process of constructing a rung has been:

1 Find a measurable quantity associated with a class of objects.

2 Observe how the measurable quantity depends on distance for objects close enough to have had their distances measured by the method of a previous rung.

3 Assume the same relationship holds for more distant objects of the same class, and hence calculate their distances.

4 Return to step 1, with a new measurable quantity.

The classes of objects (and distance indicators) for the first four rungs of the distance ladder were:

Sun	(by radar ranging)
Nearby stars in our Galaxy	(by triangulation)
Our Galaxy and nearby galaxies	(using Cepheid variables)
Nearby and somewhat further off galaxies	(using Type Ia supernovae)

These distance indicators all depended on recognizing a particular type of *star*. But, as was mentioned in Section 3.1, individual stars can be resolved only in galaxies that are not too distant. For most galaxies, a method is needed which depends on recognizing, or deducing, the luminosity of the galaxy as a whole. Although, as already noted, individual galaxies vary considerably in their luminosity, they occur in clusters. A simple rule which seems to work in practice is to assume that the third-brightest galaxy in all clusters has the same luminosity (a standard 10^{37} W lamp).

Figure 9 Spiral galaxy M81 (NGC 3031), taken with the 200-inch telescope at the Palomar Observatory in California. This galaxy, like our own, has tightly wound arms and a prominent bulge (known as a nuclear bulge).

An alternative method is to separate galaxies into different types, with the assumption, or at least the hope, that the types have characteristic luminosities. There are certainly generic differences between spiral galaxies (Figure 9) and elliptical galaxies (Figure 10). But it is also well established that there is a useful correlation between the luminosities of spiral galaxies and their rotation speeds, which can be determined from (radio) observations of Doppler broadening.

Figure 10   M59 (NGC 4261), an elliptical galaxy about 19 Mly (i.e. 1.9×10^7 ly) away. Like all elliptical galaxies, M59 has no spiral arms.

size of radio galaxies

Another method exploits the fact that some types of *radio galaxy* (so-called because they are strong emitters at radio frequencies) are fairly uniform in *size*, and radio interferometers can resolve very small angular separations — a thousand times smaller than those resolved by optical telescopes. The apparent size leads to an estimate of the distance of the galaxy. Because some radio galaxies are also visible at optical wavelengths, the optical and radio distance scales can be intercalibrated.

Yet another method should also be mentioned. It involves the behaviour of a single star in a galaxy, though one too distant to be resolved. We have already mentioned Type Ia supernovae. These are events that occur when a white dwarf (which itself originally resulted from the last stages in the active life of a medium-sized star) captures material from a companion star, and undergoes collapse to a neutron star. If instead of a medium-sized star one begins with a very massive star, then at the end of its active life it catastrophically collapses *directly* to a neutron star or black hole. This leads to a gigantic explosion even more energetic than a Type Ia

Type II supernova

supernova; it is called a *Type II supernova*. These explosions are bright enough to be visible at large distances — in some cases briefly shining more brightly than all the other stars of the galaxy put together. Unfortunately they are rather rare, occurring only once every hundred years or so in a typical galaxy. A Type II supernova explosion causes a spherical shell of hot gas to expand out of the star at high speed — thousands of kilometres per second. The spectral lines in the observed light from this shell (mostly from hydrogen and therefore easily identified) are *blueshifted* by its velocity towards us. Knowing this velocity from the amount of blueshift, the increase in size of the shell, month by month, can be calculated. Thus the shell is a source of known size, even though this cannot be resolved from an observed angular width. The temperature can be found from the overall shape of the continuous spectrum between the spectral lines. Knowing both the size and temperature of the shell, its total light output (that is, its luminosity) can be found. From this and the observed flux density, the distance can be calculated using the inverse square law.

Remember, luminosity $= 4\pi r^2 \times$ flux density, where r is the distance.

Figure 11   NGC 5457, a spiral galaxy in our Local Group, having looser arms and a far less noticeable nuclear bulge than the galaxy shown in Figure 9.

Objective 3   **SAQ 5**   Comment on the truth or otherwise of the following statements:

(i) Averaging statistically over the luminosities of its constituent galaxies, each cluster of galaxies can be assumed to have the same overall luminosity.

(ii) The distance-measuring method involving Type II supernovae relies on the fact that the shell of material thrown out by the explosion greatly exceeds the parent star in size, to the extent that it can be optically resolved.

4 The variation of redshift with distance

4.1 Hubble's discoveries

In this Section, we bring together two important features of galaxies — their redshifts and their distances.

This crucial development owes its origins to Edwin Hubble. His pioneering work in 1923 first led to the confirmation that certain of the fuzzy patches in the sky, loosely called 'nebulae', were in fact galaxies like our own.

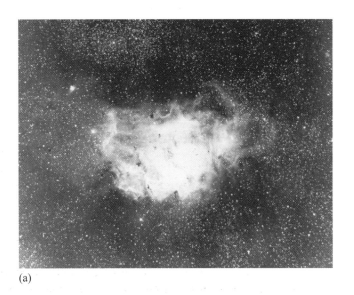

(a)

(b)

Figure 12   (a) Nebula NGC 6514, a cloud of gas and dust in our own Galaxy.

(b) Hubble showed that objects like M31 (NGC 224, now called the Andromeda galaxy) were galaxies like our own, and they ceased to be called nebulae. The two large bright patches near M31 are satellite galaxies, NGC 205 and NGC 221.

It was immediately realized that the Universe was enormously bigger than had previously been thought. Also the number of galaxies was large. In fact, it is now known that the number of galaxies accessible to our telescopes is comparable with the number of stars in our Galaxy — about 100 billion (i.e. 10^{11}).

A second significant discovery made by Hubble concerned the spectra of galaxies, nearly all of which are redshifted. This redshift was a systematic shift of *all* the lines to the red end of the spectrum (it was discovered by another US astronomer, Vesto Slipher). Using his measurements of distance, Hubble showed that the *redshift increased with distance*. As far as he could tell, the redshift of a galaxy was proportional to its distance.

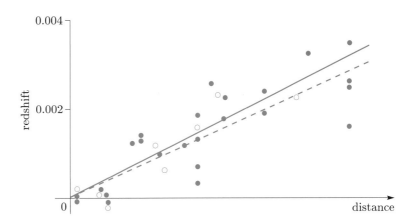

Figure 13   The redshift–distance relationship for galaxies, as plotted by Hubble in 1929. The solid line represents the relationship inferred from individual galaxies (solid circles), the dashed line the relationship when the galaxies are combined into groups (open circles). Hubble's distance scale has been omitted since it is now known that it was in error.

Hubble's original measurements, shown in Figure 13, exhibited a large scatter about a straight line. This was partly due to the inevitable observational uncertainties, especially in distance measurements. But even if one were able to remove the observational uncertainties, there would still have been a considerable scatter about a straight line. This comes from the fact that most of the galaxies Hubble looked at were members of clusters, and each was moving about within its cluster. Because the galaxies were rather close to us, the speeds associated with this random motion were comparable to the recessional speed Hubble was trying to measure. Sometimes the motion of the galaxy relative to the cluster centre was directed towards us (giving a blueshift component), sometimes away from us (adding a further redshift component). If the structure of the cluster were well enough known, these effects could, in principle, be estimated and a correction applied. Allowance must also be made for small gravitational redshifts (see SAQ 2). Finally, the motion of the Earth with respect to our Local Group must be subtracted. (This will be discussed in Section 6.2.4.) Suppose that *all* these corrections could be perfectly made in all cases. Then a redshift (z) against distance (r) plot would, as far as we know, appear as in Figure 14.

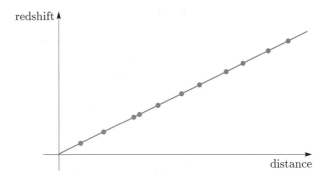

Figure 14   An idealized Hubble diagram with all sources of scatter removed.

The straight line in Figure 14 would represent the underlying cosmological

cosmological redshift   redshift. Thus, for this *cosmological redshift*, we have

$$z \propto r. \tag{3}$$

Recall that the suffix 0 refers
to the value at the *present*
time, while the suffix 1 refers
to the value at the *earlier*
time.

Interpreting the observed redshift as a Doppler shift implies that each
galaxy is receding at a speed proportional to its distance from us. To see
this, consider the Doppler shift formula given in Equation 5 of Unit 6
(setting $f = f_1$ and $f' = f_0$), and the relationship between redshift,
wavelength change and frequency shift given in the answer to SAQ 3 of
this Unit. It follows that the redshift, z, is given by

$$z = \frac{\lambda_0 - \lambda_1}{\lambda_1} \approx \frac{\lambda_0 - \lambda_1}{\lambda_0} = 1 - \frac{\lambda_1}{\lambda_0} = 1 - \frac{f_0}{f_1} = 1 - \sqrt{\frac{1 - (v/c)}{1 + (v/c)}} \approx \frac{v}{c}. \quad (4)$$

Hence, Equations 3 and 4 together imply that

$$v \propto r. \qquad (5)$$

Hubble's law

This provides the basis for one of the common ways of writing *Hubble's
law*,

$$v = Hr. \qquad (6)$$

Hubble parameter, H

The factor of proportionality, H, is sometimes called the *Hubble constant*.
But the term *parameter* is perhaps preferable since the word 'constant'
might lead one to think that it should remain constant in time — instead
of being a constant of proportionality between two variables *as those
variables are at a particular point in time*. H must, we now realize, vary
slowly with time.

If the speed of recession is proportional to distance, this implies *all*
distances between galaxies are increasing at the same rate. Not only do all
clusters of galaxies appear to be receding from us here on Earth, they
would appear to be receding in exactly the same manner from whatever
vantage point an observer adopted.

Hubble's work has been continued, refined, and extended to much more
distant and fainter galaxies.

Objectives 4 and 5   **SAQ 6**   Taking typical intercluster distances to be approximately 2×10^8 ly, and
the value of H to be about $2 \times 10^{-18} \, \text{s}^{-1}$, estimate the minimum z value
that can reliably be ascribed to the expansion of the Universe. Compare
this with the redshifts measured by Hubble in Figure 13.

Since it is hard to measure the distance to a far-off galaxy, it is not
surprising that there has been a good deal of controversy about the
reliability of distance estimates. The redshift measurements are much
easier to make and are more direct, so there has been much less
uncertainty over them. Nevertheless, it is useful to have a check, and this
has been provided by radio astronomers measuring the redshift of lines in
the radio spectrum, such as that due to the emission of hydrogen at a
wavelength of 21 cm. This confirms that the redshift is the same over a
large range of frequencies.

Based on a range of recent results, the value of H is currently reckoned to
be $2.3 \times 10^{-18} \, \text{s}^{-1}$, with an observational uncertainty of about 10%.

It has become common usage to write
$H = h \times 100 \, \mathrm{km \, s^{-1} \, Mpc^{-1}}$,
in which case h currently has a best estimate of 0.7.

Hubble, who interpreted redshifts in terms of recessional velocities, would quote this as $23 \, \mathrm{km \, s^{-1}}$ for every million light-years of distance of a galaxy. Estimates of H are often quoted in terms of $\mathrm{km \, s^{-1} \, Mpc^{-1}}$. In these units, H is about $70 \, \mathrm{km \, s^{-1} \, Mpc^{-1}}$.

We end this Section with a more up-to-date version of the Hubble diagram (Figure 15) for some galaxies and clusters of galaxies.

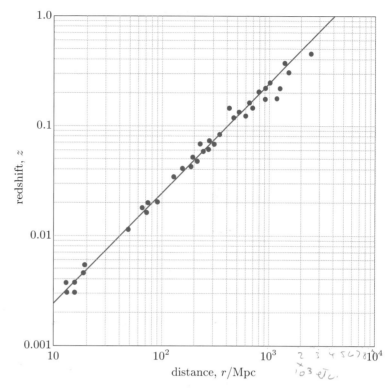

Figure 15 A plot of redshift against distance for a selection of galaxies and clusters of galaxies.

Objectives 4 and 5   **SAQ 7**   Use the sum of the shifts calculated for the star specified in SAQ 2 to give an estimate of the uncertainty in the cosmological redshift z for a star in a galaxy at $4 \times 10^6 \, \mathrm{ly}$ from us. (This may be an overestimate because the centre of a line can be estimated quite well even if the line is fuzzy. However, in a real galactic spectrum there will be a further component of broadening due to galactic rotation.)

Objective 4   **SAQ 8**   By taking appropriate readings off the graph of Figure 15, estimate the value of the Hubble constant, H, indicated by this set of data.

4.2 Evidence for a big bang

Having interpreted the redshift as indicating a recessional speed proportional to distance, one may extrapolate into the future to predict how the positions of the galaxies will evolve with time. One can also run the sequence backwards, so to speak, to discuss what their positions were in the past. Clearly, at former times the galaxies were closer to each other.

big bang

But not only that. Because of the proportional relationship between speed and distance (Equation 6), at a certain time in the past, all the matter of the Universe must have been together at a point of extraordinarily high density. It was from this condition that it subsequently expanded giving the matter of the Universe its present-day distribution. This is our first indication that the history of the Universe featured an explosive event. This has become known as the *big bang*. It is believed to have marked the beginning of the Universe. (Actually the phrase 'big bang' is used in two ways: (i) to denote the instant at which cosmic expansion begins; and (ii) to refer to that instant plus the sequence of events immediately following. It is usually clear from the context which of these meanings is intended.)

In the next Units, we shall have much to say about the nature of the big bang and how long ago it took place — in other words, how old the Universe is. But before coming to that we need to be sure that there really *was* a big bang. What we seek is evidence that is independent of the observation of moving galaxies. The remainder of this Unit is devoted to describing just such confirmatory observations. Not only do they add to our confidence that the Universe did indeed have a definite beginning, they also inform us that the beginning was exceedingly violent — the big bang was hot. This indication assumes great importance later when we seek to get some understanding of the varying types of process that must have been taking place during the initial stages of expansion — during the first years, minutes, even fractions of a second after the instant of the big bang.

5 The microwave background radiation

5.1 A second major discovery

In the introduction to this Unit, we said that there were three pillars of evidence for the big bang. We now turn to the second. It rests on a discovery that ranks in importance with that of Hubble's law. It came about when observations in a new region of the electromagnetic spectrum — the microwave region — became possible. This was due to the invention of new detectors, working at frequencies as high as 30 000 MHz. In 1965, two Bell Telephone scientists, A. Penzias and R. Wilson, were investigating the radio noise found at wavelengths between a few millimetres and a few centimetres. These wavelengths were, at the time, a relatively untapped field for communications. (They are now very useful for satellites because even small antennae give narrow beams at these wavelengths. Penzias and Wilson were working on the Telstar/Echo satellite project at that time.)

They found that, once all known sources of noise had been accounted for, they were left with a residual signal which was coming equally from all directions. It was soon realized that because of this isotropy, it could not originate on the Earth, or in the Solar System. Nor could it be coming even from the Galaxy — the Galaxy being a thin disc, with us not at its centre.

Objective 8

SAQ 9 Consider our Galaxy to be a uniform disc which is generating radio waves uniformly throughout its volume (Figure 16). Assume for the purposes of this question that all other radio sources are negligible. Also assume that the Earth is in the central plane of the disc but off centre (nearer the edge than the centre) and that it is at rest in the Galaxy. *Sketch* graphs showing the way the signal picked up by a radio telescope on Earth will vary when the telescope rotates:

(a) in the plane of the Galaxy;

(b) in a plane perpendicular to the Galaxy.

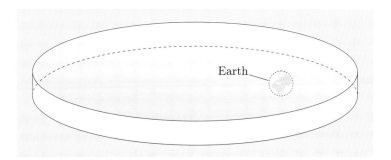

Figure 16 The Earth inside the Galaxy, shown schematically as a disc.

Having separated out these other sources of background noise, it was concluded that the isotropic component had to be of cosmic origin. It is called the *cosmic microwave background radiation*.

cosmic microwave background radiation

Figure 17(a) shows how complicated the overall microwave spectrum is, due largely to a set of lines generated in the Earth's atmosphere. Figure 17(b) focuses on the region at the left-hand side of Figure 17(a). This region is of interest because it is at lower frequencies than most of the atmospheric lines. The atmospheric interference varies with the thickness of the atmosphere, and hence with the angle of observation. This noise can therefore be separated out from the cosmic signal in which we are interested. Figure 17(b) shows the microwave spectrum after correcting for the effects of the atmosphere.

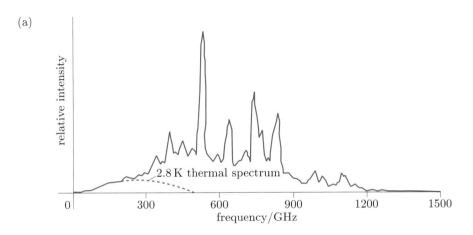

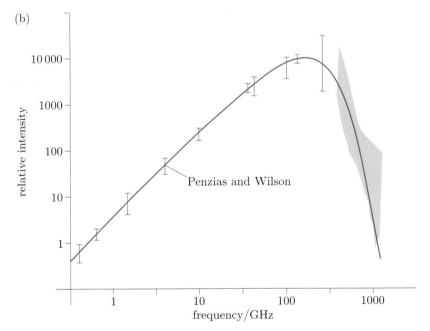

Figure 17   (a) The overall ground-based microwave spectrum.

(b) Ground-based measurements by Penzias and Wilson, and by other observers, of the microwave intensity at various frequencies; an enlargement of the spectrum shown at the left-hand side of (a), corrected to remove the effects of the atmosphere. The shaded area represents the limits given by measurements made with a detector that covered a wide band of frequencies. (The solid line shows a thermal spectrum corresponding to a temperature of 2.8 K. The term 'thermal spectrum' is explained a little later in the text.)

In 1989, the COBE satellite was launched. It was able to make measurements from above the Earth's atmosphere and was therefore not subject to some of the problems encountered by Penzias and Wilson, and by other ground-based observers. However, a major contaminant —

radiation originating from within the Galaxy — remained. The COBE results, after correction for this effect, are shown in Figure 18.

The shape revealed by this closer look is identical to that found for the spectrum inside a hot cavity — for example the spectrum in an oven or furnace (Figure 19). It is called a 'black body' spectrum (because it is that which is given out by an idealized heated black surface), or simply a *'thermal' spectrum.*

thermal spectrum

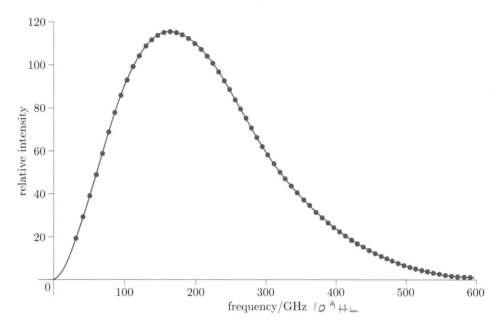

Figure 18 Data on the microwave background radiation taken by the COBE satellite. The curve through the data points is that of a thermal spectrum of 2.73 K. Note that Figure 17(b) had a logarithmic frequency scale, while this one is linear.

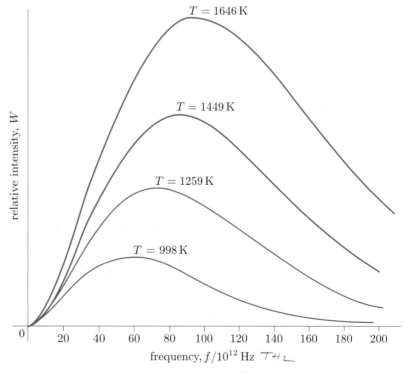

Figure 19 Thermal spectra for various temperatures, T, based on laboratory measurements of relative intensity, W, of radiation with frequency f.

One of the basic features of thermal radiation is that, regardless of the temperature of the surface or enclosure generating it, the *shape* of the spectrum is always the same. The *peak* of the spectrum moves up to higher frequencies as the temperature rises (it might glow red hot or white hot). But the basic shape remains the same. What this means is that if the spectra for two different temperatures are drawn on two graphs, we can always choose scales — linear but different — so that the graphs can be superimposed (Figures 20 and 21). This single, characteristic shape arises from, and is an indicator of, the equilibrium conditions inside the oven. The rescaling in Figure 20 works *only* for thermal spectra, and so is a true indicator of equilibrium.

As an example of the opposite extreme, take the line spectra from a hydrogen lamp and a sodium lamp. No amount of rescaling would fit these two together; they are quite distinct.

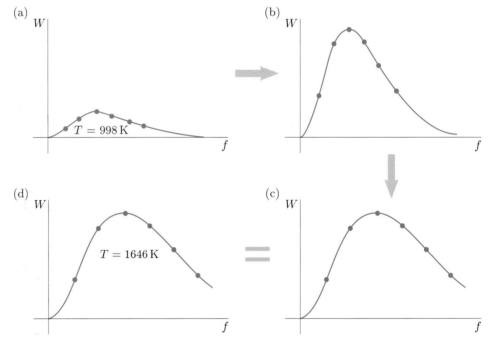

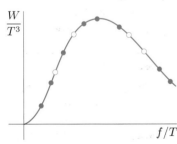

Figure 21 As suggested by Figure 20, a plot of W/T^3 as a function of f/T is a single curve, for all temperatures; that is, if W/T^3 is plotted against f/T for a number of frequencies and temperatures, the plotted points for all frequencies and temperatures lie on one and the same curve.

Figure 20 What is meant by the shapes of thermal spectra being independent of temperature.

(a) is the measured spectrum at 998 K (from Figure 19);

(b) is obtained from (a) by expanding the vertical scale by a factor of $(1646/998)^3$;

(c) is obtained from (b) by expanding the horizontal scale by a factor of $(1646/998)$;

(d) is the measured spectrum at 1646 K (from Figure 19) plotted using the same scales as those in (a), and is the same curve as (c).

Since the cosmic microwave spectrum has a thermal shape; the conclusion is that it was generated in equilibrium conditions — in some sort of cavity. But the radiation fills the Universe: the 'cavity' is the Universe itself, a cavity with no walls.

The word 'equilibrium' is used here with some reservations. The Universe has not reached overall equilibrium, and strictly speaking never will. This is because of the expansion of the Universe. We saw in Section 4.1 how the wavelength of light from distant galaxies is redshifted due to the expansion of the Universe. It turns out that the same thing happens to the microwave background radiation; its wavelength is also increased. The peak frequency of the spectrum is reduced, and this means the radiation is progressively cooling. Hence the radiation has not strictly speaking reached equilibrium. However, the processes which transferred energy between radiation and matter in the early Universe were so rapid that it is meaningful to think of a quasi-equilibrium state having been attained at an early time, with the temperature gradually falling subsequently because of the expansion of the Universe.

A second feature of the thermal spectrum is that if the detector is situated within the 'oven' generating it (as distinct from looking at a distant black surface or opening to an oven), the intensity of the radiation at any particular frequency uniquely identifies the thermal curve to which it belongs, i.e. what the temperature of the 'oven' is. So, if Figure 19 referred to thermal spectra picked up by a detector situated *inside* an oven, the ordinate of the graph could be expressed in terms of *absolute*, rather than *relative* intensities. Under these circumstances, a measurement of intensity at a single frequency, say 60×10^{12} Hz, would be sufficient to identify which curve that data point belonged to, and hence what the temperature of the oven was.

3 K radiation

Inasmuch as the Universe can be regarded as an 'oven', and we are in it, Penzias and Wilson were able to estimate from the intensity of the radiation at the single frequency they were detecting that the temperature was about 3 K. For this reason, the cosmic microwave background radiation is often called the *3 K radiation*. The spectrum observed by COBE allowed a more precise estimate of the temperature, namely 2.73 K. This is based on the results presented in Figure 18, where it should be noted that the data points do not deviate from the thermal curve by more than 0.03%, this being consistent with measurement precision.

In the 1940s, some theoreticians already had an inkling that this radiation should exist, from the predictions of their cosmological models. The Abbé Georges Lemaître, a Belgian cosmologist, was the first to see clearly (around 1927) that the expansion of the Universe pointed back to a 'big bang'. But he could not get much further because not enough was known about nuclear physics at that time. It was George Gamow and his colleagues, in 1948, who first saw that very high temperatures must be involved at early times in an expanding Universe. They sketched out some nuclear reactions that must therefore have taken place. By 1953, their reconstruction had been refined and gave definite predictions for nuclear abundances — and radiation intensity. It was at this stage they realized that radiation had been a vital component in the early Universe and that this same radiation, albeit substantially redshifted and cooled, should still be around today (there being no other place for it to go!). In the 1950s, radio equipment already existed which was sensitive enough to detect this radiation. Indeed, radio astronomy was developing fast, based at first on the technology of military radar in World War II. So by 1953 the stage was set for this radiation to be discovered. But as it so happened, the two groups, theoretical and experimental, did not stumble into each other for another 12 years. Despite conferences and journals, scientific communication with New Jersey was not as effective as communication over 10^{10} light-years! When theory and observation finally came together, the question of priority took some sorting out, as indicated in Gamow's letter, reproduced as Figure 22.

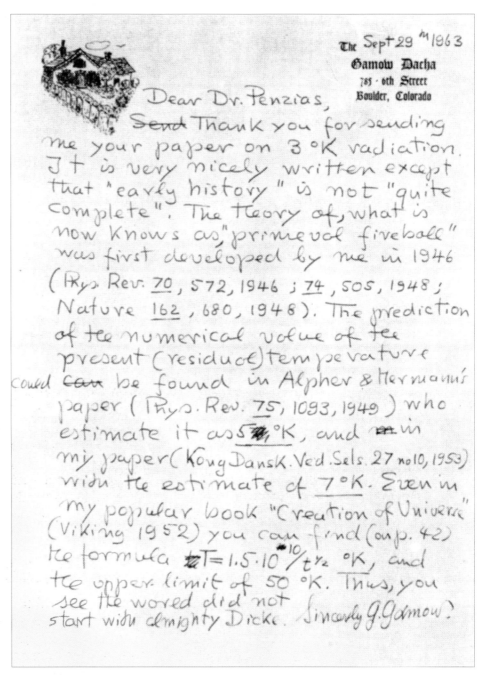

Figure 22   A letter from George Gamow to Arno Penzias.

The reference to 'almighty Dicke' at the end of this letter concerns
R. H. Dicke, the leader of a group at Princeton University (also in New
Jersey) which was pursuing both theoretical and observational research
into the background radiation in the 1960s. The paper by Penzias and
Wilson announcing their discovery, and a paper by Dicke's group providing
a possible cosmological interpretation, were published back to back in
volume 142 of the *Astrophysical Journal* in 1965.

The spectrum of thermal radiation at a given temperature T can be
expressed in terms of a function $W(f, T)$ which gives the intensity of
radiation of frequency f. The energy density of that part of the radiation
with frequencies lying between f and $f + \Delta f$ is given by $W(f, T)\,\Delta f$. The
formula for $W(f, T)$ was derived by Max Planck in 1900:

$$W(f, T) = \frac{A f^3}{\exp(Bf/T) - 1} \tag{7}$$

where A and B are constants. At low frequencies ($f \ll T/B$), one can use a simpler approximate formula:

$$W(f, T) \approx \frac{A}{B} f^2 T. \tag{8}$$

Study comment

We suggest that you do not omit the following SAQ as it contains information that will be needed later.

Objective 6   **SAQ 10**   (a) Verify the result represented by Figure 21, that W/T^3 is a function of f/T only, which applies for all frequencies and temperatures.

(b) Use the result of (a) to show that maximum intensity occurs at a frequency, f_{max}, which is proportional to the temperature.

(c) Imagine that you receive a radio message from a very distant galaxy, informing you about measurements of the cosmic background radiation at various frequencies (defined in terms of fractions of the frequency of a standard spectral line). These results do not agree with your own measurements of the radiation. In particular, you find that the quoted value of the maximum intensity is *eight* times your value and occurs at a frequency that is *twice* your frequency for maximum intensity. You find that the quoted intensities at low frequencies are *twice* your values at the same frequencies. How can you explain these discrepancies? (You may assume that the discrepancy is not due to calibration or other errors.)

Objective 6   **SAQ 11**   (a) On the basis of your solutions to parts (a) and (b) of the previous SAQ, which would be hotter, a red star or a yellow star?

(b) The curves of Figure 19 can be used to extrapolate results to lower temperatures. By taking a measurement off the figure, and using the result of SAQ 10(b), estimate the frequency of the intensity maximum for radiation emitted by a body at room temperature. What is the term used to describe electromagnetic radiation in this frequency range?

5.2   The origin of the 3 K radiation

In speaking of the radiation as having a cosmic origin, what do we have in mind? Essentially this:

In the violent conditions of the early evolution of the Universe, a stage was reached where the matter consisted of a plasma of electrons, protons, neutrons, and some light nuclei such as helium. There were no atoms as such for the simple reason that atoms would have been too fragile to withstand the violence of the collisions that were taking place at the temperature that then existed. As electromagnetic radiation passed through the plasma, it interacted with the matter, exchanging energy in packets or 'quanta' of magnitude

$$E = hf \tag{9}$$

where f is the frequency of the radiation and h is the Planck constant.

The radiation was mainly affected by its collisions with the *electrons*. This is because such collisions cause much bigger energy changes to the photons than collisions with the far more massive nucleons (just as a table tennis ball may lose all its energy in a collision with another table tennis ball, but will bounce off a relatively massive billiard ball with little change in energy). Thus there is a ready exchange of energy between the photons and the electrons, in the process of which, the radiation acquires the thermal spectrum characterized by the temperature of the electrons. The radiation and the electrons tend to come into thermal equilibrium with each other, and the electrons are said to have *thermalized* the radiation.

thermalization

As the expansion of the Universe proceeded, the temperature of the radiation progressively fell, and so did that of the matter. This fall led to important changes in behaviour. From the earliest times, the Universe had been opaque to radiation, in the sense that it could not travel far before it interacted with the electrons. But as the temperature declined and photon energies decreased, a stage was reached where electrons could be bound to nuclei to form neutral atoms — atoms that were no longer likely to be disrupted in collisions with the reduced-energy photons. Later, the energy of the radiation reduced still further to the point where it could not even excite the atomic electrons to higher energy states. At this stage, the radiation could no longer be strongly absorbed by matter. This being so, the Universe became transparent. This stage we call the *decoupling* of radiation from matter. (You will find that some books refer to this stage in the development of the Universe as the 'recombination' era rather than the decoupling epoch.) It occurs when the radiation has cooled down to the point where the most probable photon energy corresponds to a temperature of 3000 K. As you will find in Unit 15, where we take you systematically through the various stages in the development of the Universe, this occurred some 4×10^5 years after the instant of the big bang. Thus the radiation we now observe as 3 K radiation is today's cooled-down remnant of that 3000 K big bang radiation.

decoupling

How confident can we be that this was indeed the origin of the 3 K radiation? There are essentially four properties that lead to this conclusion:

1 As we have already mentioned, the *isotropy* of the radiation points to some global, cosmic origin.

2 The *spectrum* of the radiation is such that it could only have been produced by a sufficiently rapid interaction of the radiation with matter for the thermal energy distribution of the particles of matter to be imprinted on the radiation. Only in the early dense stages of the Universe were particles and radiation interacting fast enough for this to have been achieved within the time available.

3 The *present temperature* of the radiation of only 3 K is lower than that of most visible matter currently in the Universe. How could it be so low? The only reasonable explanation is that it has been strongly redshifted — indicating that it has been travelling towards us over an exceedingly long period of time, i.e. it was emitted soon after the big bang.

4 The *density of photons* corresponding to the 3 K radiation is enormous. In fact there are believed to be about 10^9 times as many 3 K photons in any large region of the Universe as there are neutrons and protons. Clearly this radiation is no mere by-product of an obscure process; it is a ubiquitous feature of the Universe. This prompts us to ask at what stage of the Universe is radiation likely to have played a dominant role? The answer has to be: the violent *early* Universe.

5.3 The redshift of the 3 K radiation

The temperature, T, of the radiation is proportional to the most probable photon energy, E, which as we have said is proportional to f, and hence inversely proportional to the wavelength λ. Thus,

$$T \propto \frac{1}{\lambda}. \tag{10}$$

According to Equation 1, we have for the redshift, z

$$z = \frac{\lambda_0 - \lambda_1}{\lambda_1}.$$

Thus,

$$1 + z = \frac{\lambda_0}{\lambda_1}. \tag{11}$$

The wavelength we observe now, λ_0, is that corresponding to $T \approx 3$ K, whereas the original wavelength, λ_1, emitted during the decoupling epoch corresponded to $T \approx 3000$ K. Hence

$$1 + z \approx \frac{3000}{3} \approx 1000$$

ie from (10) $T_0 \propto \frac{1}{\lambda_0}$, $T_1 \propto \frac{1}{\lambda_1}$,

so $z \approx \frac{T_1}{T_0}$

and hence $z \approx 1000$. This compares with $z \approx 6$ for the furthest optical object so far seen (2004). This was associated with a quasar. From these z values, you can appreciate how much further back in time and farther away in distance it is possible to look with a microwave detector than with an optical telescope. Even so, it is important to recognize that we cannot, and never shall be able to, see right back to the instant of the big bang. For the first 4×10^5 years, the Universe was opaque.

Quasars are believed to be the highly luminous centres of certain galaxies at an early stage in their life. They can be 1000 times brighter than a typical galaxy and can therefore be seen at great distances.

Objective 6   SAQ 12   In Unit 15, we shall see that the wavelengths associated with photons increase as $t^{2/3}$, where t is the time they were emitted after the big bang. Assuming the time now, $t_0 = 1.4 \times 10^{10}$ years, and that decoupling took place when $T = 3000$ K, estimate the time, t_d, at which decoupling occurred.

6   The angular distribution of the 3 K radiation

6.1   Basic isotropy

angular distribution

As we have said, the photons in the 3 K background have been practically free from interaction with anything since about 4×10^5 years after the instant of the big bang. The present *angular distribution* of the microwave radiation — the way in which it is spread across the sky — is therefore almost the same as it was then. The spectrum we find today depends on the temperatures at that time — for the intensity of the radiation in a particular region of the early Universe depended *only* on the temperature. If the Universe at that time was inhomogeneous (in the sense of being hotter in some regions than others) the 3 K radiation observed today would be more intense in some directions than in others.

In fact, as we have indicated, the radiation is largely isotropic. There are, however, some small — some *very* small — but significant departures from perfect isotropy. But before these can be identified, we have to take into account the effect of the Earth's motion. This would in any event impose an apparent asymmetry on the radiation. Indeed, as we shall now show, this effect can be turned to our advantage and provide us with a means for measuring the speed of the Earth relative to the average distribution of matter and energy in the Universe.

6.2   The effect of the motion of the Earth

6.2.1   The need for a reference frame for describing the Universe

The speed of the Earth in its orbit round the Sun is $29.8 \, \mathrm{km \, s^{-1}}$, in a heliocentric frame. But to specify the velocity vector, it is not sufficient to specify the Sun as the origin of the coordinate system; fixed directions must also be identified.

Objective 6

SAQ 13   Here are two possible rules for fixing an x^1-direction:

(i) The x^1-axis is taken as the line from the Sun to the nearest star (Proxima Centauri);

(ii) The x^1-axis is taken as pointing in the direction from the centre of the Galaxy to the nearest large galaxy (the Andromeda galaxy).

Are these two methods equally acceptable? If not, which is preferable, and is it completely satisfactory?

The considerations of SAQ 13 force one to look at bigger and bigger aggregates of matter in the search for reference bodies with respect to which a system of coordinates can be defined independently of the motion caused by the gravitational effects of nearby matter in our local region of the Universe. We assume that the clusters of galaxies fulfil this role. But clusters may be loosely associated (though not bound) in superclusters.

For this reason, we would like to have an alternative way of establishing a reference frame. The 3 K radiation provides such a means.

If the 3 K radiation filled the Universe at early times and has not interacted appreciably with matter since decoupling, then this radiation defines a system in which matter and energy in the early Universe are assumed to have been distributed homogeneously. So if we were to find that the 3 K radiation is *completely* isotropic when observed on Earth, we could conclude that we are *at rest* with respect to the average distribution of matter and energy in the Universe. In that sense, the radiation can be said to define a 'rest frame' of the Universe. But if we find that the radiation is not isotropic, and moreover, that it varies in a characteristically systematic fashion according to direction, then we can conclude that we are moving in a particular direction with respect to the frame in which the radiation is isotropic.

It is a matter of great interest to know the velocities of the Earth, the Sun, our Galaxy and our Local Group with respect to the system defined by the 3 K radiation. It might, for example, provide support for the idea that gravitational effects within a supercluster must be taken into account in mapping out the Universe.

6.2.2   The Earth's motion relative to the 3 K radiation

Eg 9, $E = hf$. Although a photon has zero rest mass, it carries momentum $h\frac{f}{c}$.

As noted in Unit 4, radiation has energy and momentum, so we can use the molecules of a fluid such as air as an analogy for the photons of radiation. A detector pointing forwards along the direction of our motion (if any) will encounter a greater number of photons than a detector pointing backwards; in other words, it will record a higher intensity of 3 K radiation. (If the detector is tuned to a narrow band of frequencies one would also have to take account of the change in observed spectrum, but the principle remains the same.)

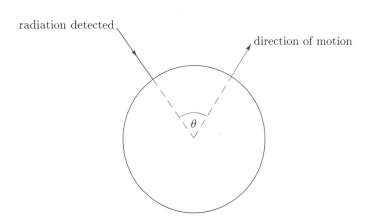

Figure 23   The radiation detected at an angle θ to the direction of motion has a thermal spectrum characterized by a temperature $T(v, \theta)$ which depends on the speed v and the angle θ.

Not only does the Earth's motion affect the intensity as a function of angle, but also the *energy* of the photons. Suppose the Earth is moving with a velocity **v** with respect to the frame in which the radiation is isotropic and we observe the component of the 3 K radiation that arrives in a direction making an angle θ with our direction of motion (see Figure 23). What sort of spectrum would we observe? One might think that the observed spectrum would be very complicated. After all, the photons have been shifted to different frequencies, because of the Doppler effect, and our clocks and measuring rods will measure different intervals of time and length, compared with clocks and measuring rods in the 'frame of isotropic 3 K radiation'. But it turns out that the result is very simple: Cosmic microwave radiation observed at an angle θ relative to the observer's direction of motion will still have a *thermal* spectrum, though it will be characterized by a temperature $T(v, \theta)$ which depends on the observer's speed v and the angle θ. (This reminds us of how the expansion of the Universe affected the cosmic background radiation: in that case too the thermal character of the spectrum was maintained; only the temperature was changed.)

It is in fact easy to calculate how $T(v, \theta)$ depends on v and θ. We shall use the fact that the frequency $f_{\max}$, corresponding to the maximum intensity, is proportional to the temperature (see SAQ 10). That means that in the 'rest frame'

$$f_{\max} = CT_0$$

where C is a constant and T_0 ($\approx 3\,\mathrm{K}$) is the temperature of the radiation in the frame in which it is isotropic. But when we are moving with a speed v and observing at an angle θ, we find a different frequency, $f'_{\max}$, corresponding to maximum intensity:

$$f'_{\max} = CT(v, \theta).$$

The relationship between $f_{\max}$ and $f'_{\max}$ is given by the formula for the relativistic Doppler shift (see SAQ 20 of Unit 7):

$$\frac{f_{\max}}{f'_{\max}} = \frac{1 - \dfrac{v}{c}\cos\theta}{\sqrt{1 - (v^2/c^2)}} = \frac{T_0}{T(v, \theta)}$$

Thus we find that

$$T(v, \theta) = \frac{T_0\sqrt{1 - (v^2/c^2)}}{1 - \dfrac{v}{c}\cos\theta}$$

$$\approx \frac{T_0}{1 - \dfrac{v}{c}\cos\theta} + \left(\text{negligible terms of order } \frac{v^2}{c^2} \text{ and higher}\right)$$

$$\approx T_0\left(1 + \frac{v}{c}\cos\theta\right) + \left(\text{negligible terms of order } \frac{v^2}{c^2} \text{ and higher}\right)(12)$$

Thus, if we were to measure the apparent temperature of the radiation at different angles, we could determine our speed v and the direction in which we are moving (which is the direction in which the radiation has the highest temperature, corresponding to $\theta = 0$).

Radio astronomers measure the intensity of radiation at a particular frequency or over a particular range of frequencies. The variation of intensity with angle is particularly simple if measurements are taken at *low* frequencies, well below the frequency for maximum intensity, $f_{\max}$. In this region of the spectrum, the intensity at a given frequency is proportional to the temperature (see Equation 8),

$$W \propto T \approx T_0\left(1 + \frac{v}{c}\cos\theta\right)$$

and hence

$$W \approx W_0\left(1 + \frac{v}{c}\cos\theta\right). \tag{13}$$

dipolar anisotropy

The prediction of Equation 13 is plotted in Figure 24 for values of $v/c = 0$, 0.1 and 0.2. The angular distributions (b) and (c) show a *dipolar anisotropy*.

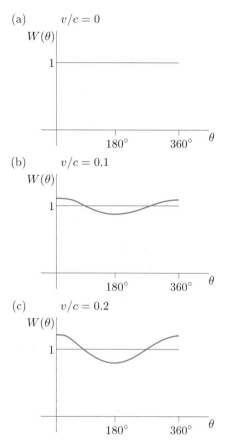

Figure 24   The radiation intensity at low frequencies, $W(\theta)$, as a function of θ, for v/c equal to (a) 0, (b) 0.1 and (c) 0.2, where $W(\theta)$ is calculated from Equation 13. (Note that v/c for the Earth's motion round the Sun is only 10^{-4}.)

At higher frequencies, the coefficient of $\cos\theta$ is more complicated and must be calculated from Equation 7. The result of such a calculation, which applies at all frequencies, is

$$W \approx W_0 \left(1 + F(f)\frac{v}{c}\cos\theta\right) \qquad \text{24 hour variation.} \qquad (14)$$

where

$$F(f) = \frac{fB/T}{1 - \exp(-fB/T)}$$
$$B = 4.8 \times 10^{-11}\,\text{K s, and}$$
$$T \approx 3\,\text{K}.$$

But for $f \ll T/B \approx 6 \times 10^{10}\,\text{Hz}$, we have $F(f) \approx 1$ and obtain the simpler result of Equation 13.

6.2.3   Measurement of the angular distribution of the 3 K radiation

How are such angular distributions to be measured? One way, of course, is to take a radio telescope and swing it round the sky, taking readings in different directions. But as is clear from Figure 17(a), the atmosphere itself emits microwaves. There is therefore a grave danger, with this method, of picking up different contributions of atmospheric emission from different directions — more as you point your telescope close to the ground, with a long path through the atmosphere, less when it is vertical with the minimum path through the atmosphere. A better, and indeed simpler, way with a ground-based telescope is to leave the telescope in a fixed direction with respect to the Earth, say vertical, and let the Earth's rotation carry it round. In 24 hours, the telescope will be carried round a whole circle, allowing the measured intensity to be plotted against angle of observation as indicated in Figure 24.

Looking at the Earth from outside the Solar System, one sees it revolving $366\frac{1}{4}$ times a year, not $365\frac{1}{4}$. From our position on the Earth, we see only $365\frac{1}{4}$ days, because one turn is taken up in the revolution round the Sun. A sidereal day is $(1/366\frac{1}{4})$ of a year and a solar day is $(1/365\frac{1}{4})$ of a year.

Thus a practical prescription for looking for the translational motion of the Earth, relative to the frame in which the 3 K radiation is isotropic, is to look for a variation with a 24-hour period in the radiation detected by a fixed radio telescope. Such a search, for a variation of *known period*, can be carried out very sensitively. That is why we have looked at the data in two stages, first asking if the data look isotropic, and now asking more specifically for the magnitude of any 24-hour variation.

A further nice experimental feature is that the component being sought is of period 24 *sidereal* hours. Human interference, on the other hand, will tend to have a period of 24 *solar* hours. This small distinction (explained in the marginal note alongside) often allows a radio astronomer to sort out the pure astronomical signal from the strong terrestrial noise.

Objective 7   **SAQ 14**   How large a change in apparent temperature of the 3 K radiation is expected due to the Earth's motion round the Sun? How large is it due to the Solar System's motion of $230\,\mathrm{km\,s^{-1}}$ round the Galaxy? (The back cover gives the information you may need to answer this question.)

6.2.4   The speed and direction of the Earth's motion

The first significant claim to have detected the motion of the Earth relative to the 'frame of isotropic 3 K radiation' came in 1977 from a group at Berkeley, California. They concluded that the Earth is moving at a speed of $(390 \pm 60)\,\mathrm{km\,s^{-1}}$, in a direction towards the constellation Leo, relative to a frame in which the 3 K radiation is isotropic. Their conclusion resulted from observations of a variation of intensity with angle of the form predicted by Equation 14, which we have called a 24-hour variation.

Now this motion of the Earth through the 3 K radiation will be the resultant of several component motions:

1 The Earth's velocity about the Sun (of magnitude $30\,\mathrm{km\,s^{-1}}$);

2 The velocity of the Sun itself about the Galactic centre (currently estimated to be of magnitude $230\,\mathrm{km\,s^{-1}}$);

3 The velocity of the Galaxy relative to the Local Group;

4 Whatever velocity the Local Group has relative to the 'frame of isotropic 3 K radiation'.

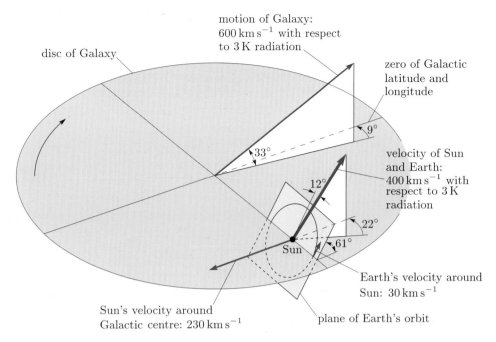

Figure 25   The absolute motion of the Earth. The Earth travels in its orbit
round the Sun at $30 \, \text{km s}^{-1}$ and is being swept around the centre of the Galaxy at
$230 \, \text{km s}^{-1}$. Experiment shows that the Earth's net speed through the 3 K
radiation is about $400 \, \text{km s}^{-1}$. The Earth's net velocity lies in the same plane as
its orbit round the Sun and at an angle tilted sharply upwards (northwards) from
the plane of the Galaxy. In this diagram, the Earth's net velocity is depicted as a
heavy arrow centred on the Sun (pointing upwards and to the right), since the two
bodies travel together. Both are being carried by the Galaxy's own motion
through the 3 K radiation. In order to account for the Earth's motion with respect
to the 3 K radiation, the Galaxy must be travelling at about $600 \, \text{km s}^{-1}$ in the
direction shown by the coloured arrow centred on the disc of the Galaxy.

The Earth's speed of about $400 \, \text{km s}^{-1}$, relative to this frame in which the
radiation is isotropic, is comparable to its speed of about $230 \, \text{km s}^{-1}$
relative to the Galactic centre of mass. Nevertheless, the Galaxy as a
whole must be moving through the 3 K radiation even *faster* than the
Earth, because the direction of the Solar System's orbital velocity round
the Galaxy is almost *opposite* to the direction of maximum observed
intensity of the 3 K radiation and hence opposite to the direction of the
Earth's velocity through the 3 K radiation, as depicted in Figure 25.
Adding the two velocity vectors gives the centre of the Galaxy a velocity
whose magnitude is about $600 \, \text{km s}^{-1}$ with respect to the 3 K radiation.
Now *if our Galaxy were isolated*, this velocity could only be interpreted as
a departure (and $600 \, \text{km s}^{-1}$ would be an embarrassingly large departure)
from the basic idea of cosmology that the expansion of the Universe is
shared by all matter and radiation; an isolated galaxy, or the centre of
mass of a cluster of galaxies, should *not* be moving with respect to the 3 K
radiation. But our Galaxy is not isolated, it is a member of our Local
Group of galaxies. It cannot be stationary with respect to the centre of
mass of the Local Group but must, to avoid falling in, be travelling around
in some quite complicated orbit. So the next step towards an
understanding of the $600 \, \text{km s}^{-1}$ is to subtract from it the velocity of our
Galaxy in its movement about the Local Group. Unfortunately this
velocity is not very well known, because the estimates of the masses of
some of the galaxies in the Local Group are very rough. But current data
give the speed of our Galaxy with respect to the Andromeda galaxy (which
is the other major member of our Local Group) as $40 \, \text{km s}^{-1}$. This might

be an indication of the kind of velocity we have relative to the centre of mass of the Local Group. If so, it still leaves an unexplained speed of $500–600\,\mathrm{km\,s^{-1}}$.

Our Local Group is thought to be a member of a cluster of clusters called the Local Supercluster. So the next step is to subtract the velocity of our Local Group with respect to the centre of mass of the Local Supercluster from the $600\,\mathrm{km\,s^{-1}}$. If this is also the final step, the answer should be compatible with zero. Unfortunately, the uncertainties are, at this stage, too great to be able to decide whether the velocity of our Local Group can be entirely explained in terms of the effects of a Local Supercluster. If not, then the effects of other structures such as more distant superclusters and the voids between them, must also be taken into account.

6.3   Anisotropies in the Universe itself

Having subtracted the dipolar anisotropy due to the motion of the Earth relative to the 3 K radiation, we are left with radiation that is exceedingly isotropic. So, we have to ask whether there are any residual variations that would point to a departure from isotropy of the radiation *itself*? This is a crucial question. Although it was gratifying to have the radiation so isotropic that there could be little doubt of its cosmic origins, nevertheless a *completely* isotropic distribution would bring troubles of its own.

The reason for this is that, although the *matter* distribution is isotropic on a large enough scale, it is clearly not so on smaller scales. It is clumped together in galaxies, the galaxies are preferentially to be found in clusters of galaxies, and even the clusters are loosely associated in superclusters. In fact, the matter distribution somewhat resembles a gigantic sponge; it has enormous holes in it, with the superclusters arranged around the boundaries of these voids. Presumably this distribution came about as a result of inhomogeneities in the original distribution of matter coming from the big bang. If a particular region happened to have by chance a somewhat greater density of matter than its neighbours, its increased gravity would tend to attract matter away from the less densely populated regions and towards itself. This would enhance the inhomogeneity, leading to this particular region gaining an even stronger pulling power, and attracting yet more material to itself. The initial inhomogeneities, which in themselves may have been slight, would over the course of time have become magnified.

That is thought to be the process whereby we have our present-day distribution. Having said that, it is not at all clear yet in what order the hierarchy of structures formed. Perhaps the matter first assembled to form superclusters; these then broke down into their component clusters, which in their turn later separated out into galaxies. Finally, the individual stars condensed. Alternatively it could all have happened the opposite way round, with stars forming first, these being attracted into galaxies, which later gathered into clusters and superclusters. Or indeed it could have been some other mix of aspects drawn from both these scenarios. But whichever was the correct sequence of events, one thing is clear: there had to be density inhomogeneities on some scale or other.

From this we infer that there should also be anisotropies in the 3 K radiation. The reason is that when gas collects together and is squashed down by its mutual gravity, potential energy is converted into kinetic energy leading to a temperature rise. It is such temperature increases that can ignite nuclear reactions and result in the birth of a new star (assuming sufficient gas has been collected originally). Gas collecting to form a primordial galaxy or a cluster of galaxies will similarly undergo a temperature rise, this rise being reflected in the type of radiation it emits. The angular distribution of this radiation, as we receive it today, should show a degree of anisotropy because it originates in matter that was not itself entirely homogeneous. Inhomogeneities must have already been present in matter when it was emitting what is now the 3 K radiation. Thus we should expect the 3 K radiation to manifest some degree of anisotropy; it should *not* be wholly uniform.

This is not to say that we should necessarily expect to detect 'hot spots'. The situation is somewhat more subtle than that. We have to recall that when we are dealing with galaxies or with clusters, a great deal of matter is involved. The gravitational potential energies are enormous. Radiation emitted from the depths of one of these conglomerations of matter has to escape the gravitational field of the matter producing it. This, as we would expect from Block 3, will lead to a gravitational redshift, i.e. a cooling of the radiation. The interesting question then becomes whether the hot radiation from the interior is still 'hot' when it escapes, or whether it will now have lost so much energy through redshift that it emerges *cooler* than the ambient temperature of the surrounding matter. We might therefore observe 'cold spots' rather than 'hot spots'.

So, although the theoretical analysis is complicated, there have to be anisotropies of one kind or another in the 3 K radiation at some level of sensitivity, and definite predictions have long been made about the angular scale on which such anisotropies ought to appear. The refinement of these predictions and the effort to detect the anisotropies observationally have become major themes in the recent development of cosmology.

There was a great stir, which even the popular press recognized, when in 1992, the COBE satellite succeeded in detecting the anisotropies, albeit at the extremely low level of 1 part in 100 000. Figure 26 shows a picture of the measured intensity distribution across the sky.

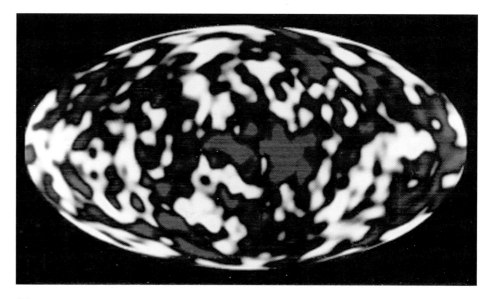

Figure 26   Departures from isotropy in the 3 K radiation (COBE satellite).

This first detection of the intrinsic anisotropies in the cosmic microwave background had to be interpreted with great care since the signal being detected was of the same order of magnitude as the background noise

angular power spectrum

fluctuations. However, since that first detection many other studies have been carried out and the angular distribution of the anisotropies has been characterized with greatly increased precision.

The characterization of the angular distribution of anisotropies is usually expressed through a plot of the *angular power spectrum* of the observed radiation. Such a plot indicates the relative strength of intensity (or temperature) fluctuations as a function of the angular scale of those fluctuations. A recent determination of this angular power spectrum, based on results obtained by COBE's successor, the Wilkinson Microwave Anisotropy Probe (WMAP), is shown in Figure 27.

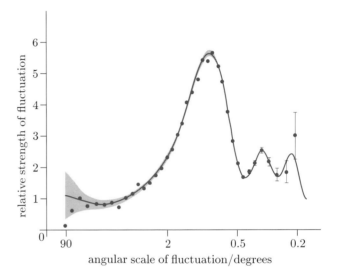

Figure 27   The angular power spectrum of the cosmic microwave background radiation as determined by the WMAP satellite (based on Bennett *et al.*, 2003).

The details of the angular power spectrum need not concern us here, but the following points should be noted.

1 The points and vertical 'error bars' represent the results of observation. The smooth line represents a 'best fit' to the data based on specific theoretical assumptions.

cosmic variance

2 The shaded band represents the unavoidable uncertainty (known as *cosmic variance* in this case) associated with trying to determine a 'cosmic' quantity from observations made at one (typical) point in the Universe, i.e. from the neighbourhood of the Earth.

3 The COBE data were limited to angular scales of 10° or more. The WMAP data reveal anisotropies on much finer scales and tell us much about the angular distribution of those anisotropies.

4 The anisotropies are particularly powerful on a scale of about 1°. This is just the angular scale at which the inhomogeneities associated with the formation of superclusters are expected to leave their imprint on the 3 K radiation.

5 Theoretical explanations for the peaks and troughs seen in the data depend on the assumed values of various cosmological parameters such as the current value of the Hubble parameter. Thus the process of 'fitting' predictions to the data provides a method of determining the values of those cosmological parameters. We shall return to this point later.

In conclusion, we can say that the extreme degree of isotropy of the 3 K radiation points to its cosmic origin in the big bang, and justifies us in regarding it as powerful confirmatory evidence that there was indeed a big bang. Nevertheless, it is perhaps the exceedingly weak inhomogeneities that will ultimately prove to be of the most lasting value for cosmology. These give us what may be a unique snapshot of a stage in the early development of galaxies/clusters/superclusters — the stage reached at the time of decoupling, 4×10^5 years after the big bang.

7 The primordial nuclear abundances

So far we have presented two pieces of evidence pointing to the occurrence of a big bang: the redshift of the galaxies (indicating the continuing expansion of the Universe), and the 3 K radiation (the remnant of the primordial radiation). But there is a third imprint such a big bang ought to have left on our present-day world. We cannot at this juncture trace out the full sequence of events following the instant of the big bang (that can only be done after we have worked through the next two Units). But regardless of the exact details of the sequence, *any* big bang scenario is likely to have passed through the stage referred to in Section 5.2 where there was a plasma of electrons, protons, and neutrons. As this cooled, nuclei (and later still, atoms) would have formed in certain proportions. Whatever these proportions were, they ought to be reflected in the relative abundances of the different elements we find in the present-day Universe. The idea therefore is to try and calculate the primordial nuclear abundances, and compare them with what exists now. If there is agreement, then that will be a third piece of evidence for our model of the big bang.

7.1   The temperature of matter and radiation

The different reactions by which neutrons and protons came together soon after the instant of the big bang to produce heavier nuclei will have proceeded at different rates according to the energies of the particles involved. The first step in calculating nuclear abundances is therefore to make some assumption about these energies.

The particles at any instant have a wide range of energies; this obviously complicates matters. Fortunately, however, it is possible to make one very important simplification. This is based on the concept of *equilibrium*. We have already touched on this topic in the context of our discussion of the origin of the 3 K radiation. Let us now look into it in a little more detail:

We use an analogy. Suppose that a large number of molecules of an ordinary gas are introduced into an insulated container. To all external appearances, the state of the gas does not change with time. It is at a fixed pressure and a fixed temperature and is said to be in equilibrium. Of course, on a microscopic scale, a great deal is happening as the molecules collide with one another. Nevertheless, the measurable properties of the gas remain effectively constant. The reason for this is that the distribution of energy among the molecules fluctuates to only a very small extent about a well-defined average pattern. Thus, although the energy of any given molecule continually changes as it collides with others, details like the most probable energy for a molecule, or the probability of a molecule having an energy greater than, say, 10^{-20} J are preserved. Now it turns out that, once the temperature of the gas has been specified, it is possible to predict the average pattern of energy distribution in the gas at equilibrium — the *temperature of matter* serves as a label for the average mixture of energies of the gas molecules. For example, the most probable energy for a molecule is proportional to the temperature, and could, in the absence of a thermometer, be used to define the temperature scale.

temperature of matter

If we now have a mixture of gas and radiation, and these are allowed to come into equilibrium with each other, not only will the particles of the gas adopt a characteristic energy distribution, but so also will the photons of the radiation. As we discussed earlier in connection with the microwave radiation, this distribution of photon energies is known as a thermal spectrum, and is rather different from the Maxwell–Boltzmann distribution for molecules. Nevertheless, the distribution can again be labelled by a single number — the *temperature of radiation*. This is defined to be proportional to the most probable energy (or frequency) of a photon. Our two temperature scales for matter and for radiation are consistent in that, whenever radiation is in equilibrium with matter, the values of the two temperatures agree.

temperature of radiation

Returning now to the early Universe, it is important to realize that the very high density of matter and radiation at that time gave rise to a frequency of collisions between protons, neutrons, electrons, and photons sufficiently high to ensure that the various components of the Universe were in almost perfect equilibrium. Thus, their respective distributions of energies could be labelled by a *single* common temperature. The electrons had the same temperature as the protons and the neutrons and the photons, because the collisions were so frequent that no part of the Universe could get out of step with the rest. This is very important because it allows us to replace the great complexity of the possible energy distributions by a single parameter — the temperature.

Strictly speaking, the various equilibria — between particles, and between particles and radiation — can never have been quite perfect because the Universe was expanding all the time, and therefore cooling rapidly. There was therefore never a steady approach to a final condition such as we associate with the behaviour of, say, a cup of coffee cooling to equilibrium with its surroundings. Indeed, as we have pointed out before, the Universe has still not reached equilibrium. Nevertheless, there was near-equilibrium in the early Universe because of the high rate of collisions.

7.2   The formation of light nuclei

It is the very high temperatures that make the early stages of the big bang relatively simple to calculate. When it comes to the formation of the first nuclei, we are looking at a temperature that has dropped to about 10^9 K, this being achieved approximately 3 minutes after the instant of the big bang. For nuclei to form, the temperature must still be high enough that charged nuclei can approach each other closely — despite the electrostatic repulsion between their positive charges — thus allowing further fusion to take place. On the other hand, if the temperature is too high, any nuclei that are formed will immediately be disrupted again by a subsequent violent collision with another particle.

The only particles taking part (to any significant extent) in the reactions at this time were neutrons, protons, electrons, and photons. In addition there were neutrinos — a very weakly interacting particle. The simplest reaction, and perhaps the most important as far as cosmology is concerned, is the fusion of a neutron (n) and a proton (p) to give a deuteron, d (i.e. a deuterium nucleus) plus a photon of sufficiently high energy to be called a gamma ray (γ):

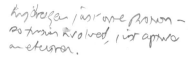

$$n + p \longrightarrow d + \gamma. \tag{15}$$

Note that the energy is mostly carried away by the gamma ray. This is essential if the neutron and proton are to bind, for otherwise the neutron and proton would simply bounce apart again. The probability of this fusion reaction occurring is high, as is that of the reverse reaction, in which a gamma ray destroys a deuteron.

As we have noted, significant numbers of deuterons could form and remain undisturbed only after the temperature had dropped to about 10^9 K. On the other hand, by the time the temperature had dropped to about 5×10^8 K, at 20 minutes after the big bang, the density had become so reduced that there were now no longer sufficient collisions taking place to produce many deuterons. Deuteron formation was therefore confined to the period during which the temperature fell between these two values.

More complicated nuclei can be produced by similar reactions. For example, helium (^{4}He) can be formed through the fusion of two deuterons:

$$d + d \longrightarrow {}_2^4He + \gamma. \tag{16}$$

But there are also several other ways in which it can be formed, each involving two steps. For example:

$$p + d \longrightarrow {}_2^3He + \gamma \tag{17}$$
$$n + {}_2^3He \longrightarrow {}_2^4He + \gamma \tag{18}$$

and

$$n + d \longrightarrow {}_1^3H + \gamma \tag{19}$$
$$p + {}_1^3H \longrightarrow {}_2^4He + \gamma \tag{20}$$

where ^{3}H is a tritium nucleus (a triton), i.e. the nuclear isotope of hydrogen consisting of a proton and two neutrons.

Each of the above reactions can occur in either direction. Thus, for example, corresponding to Reaction 20 there will be its reverse:

$$^4He + \gamma \longrightarrow p + {}_1^3H. \tag{20a}$$

Given sufficient time, an equilibrium state would arise whereby the particles present will yield the same rate for Reaction 20a as for Reaction 20. And the same will apply for each of the other reactions. Thus, corresponding to any particular temperature (the temperature governing the reaction rates) it is possible to estimate the proportions of the various particles that would be present once equilibrium had been established. This estimate is independent of whether equilibrium was reached quickly (under conditions of high density where the interactions happen frequently) or slowly (under low-density conditions).

So much for the density-independent equilibrium state achieved over a lengthy period of time. But the case of the big bang is different. The matter was rapidly dispersed only a short time after the mixture of neutrons and protons had started the chain of fusion processes. Under these circumstances, the final abundances of nuclides depended on how many collisions were able to take place before the dispersion effectively brought the processes to a halt. This in turn means that the final mix of particles *is* expected to depend on the density. More specifically, it is encounters with protons and neutrons that are significant, so it is the cosmic density of protons, neutrons and matter based on protons and neutrons that is important. This particular contribution to the overall **baryonic density** density of the Universe is known as the *baryonic density*, since the proton and neutron are the lightest members of a family of particles known as **baryons** *baryons*. In keeping with the convention introduced in Unit 10&11, we shall represent the average mass density of baryonic matter in the Universe as ρ_b/c^2, preserving the symbol ρ_b for the associated energy density of

[margin handwritten note:] Using convention ^A_ZHe, where A = mass number (protons & neutrons) Z = atomic number (protons) although not showing Z.

Roughly speaking, you can think of the baryonic density as the density of 'ordinary' matter. You are essentially made of baryonic matter as are the Earth, the Sun and all familiar objects.

baryonic matter. Thus, the baryonic density prevailing at the time of nuclear synthesis had an important part to play in governing the relative abundances of the light elements formed.

There is a second reason why the baryonic density was important. It arises from the fact that while these fusion reactions were taking place, free neutrons (those that had not yet been incorporated into nuclei) were decaying:

$$n \longrightarrow p + e^- + \bar{\nu}$$

where $\bar{\nu}$ is an antineutrino. and e^- = electron.

Unlike the fusion reactions, the rate for this decay is independent of baryonic density. Thus when the baryonic density changes, the balance between the two types of reaction changes. At low densities, a neutron travels further before colliding with another baryon, so it has a *greater* chance of decaying before being captured into a nucleus. Contrast this with collision reactions which have *less* probability of occurring at low densities. It is these different dependences on density that provide the second reason why the final mix of nuclei will depend on the baryonic density during the period of nuclear synthesis.

Now let us take a look at the result of detailed calculations. The rate of each reaction depends on the concentrations of the parent nuclei, on experimentally determined relationships between the reaction probability and the energies of the particles, and on the relationship between the equilibrium distribution of energies and the temperature. To find out the net effect of all the reactions is, mathematically, simply a matter of solving simultaneous differential equations; but they must be solved numerically, and judgement must be exercised in interpolating the experimental data. It is a lengthy computer calculation, even though it is basically straightforward. Several groups have made these calculations. We show the results of one particular investigation in Figure 28.

Figure 28 is based on a model with an assumed value of the baryonic density as it is today, $\rho_{b,now}$, from which one can extrapolate back in time to what it would have been at the time of the primordial nuclear synthesis (making due allowance for the expansion of the Universe between those two times). The *mass fractions* of the various nuclei were calculated as a function of time. Since the temperature at a given time can be calculated, the horizontal scales are marked with temperature (at the top) as well as time elapsed since the big bang (at the bottom).

mass fraction

Look first at the basic particles: neutrons and protons. Initially, at time 1 second, we have essentially 13% neutrons and 87% protons. This ratio can be calculated with confidence as it depends only on the mass difference between neutrons and protons. (We must postpone to Unit 15 a discussion of the processes occurring at epochs earlier than that of nuclear synthesis, during which this mix of neutrons and protons was established.)

As far as we are concerned, the interesting action starts just above 10^9 K, the temperature at which the proton and neutron numbers begin to drop because they are being used up in thermonuclear reactions (similar numbers of neutrons and protons are used up — it is only the logarithmic scale that makes the drop in the proton curve almost invisible). Some elements, such as helium (both the normal ^{4}He and its isotope ^{3}He), show no decrease, with the main production close to 10^9 K. The deuterium concentration, on the other hand, increases rapidly at temperatures around

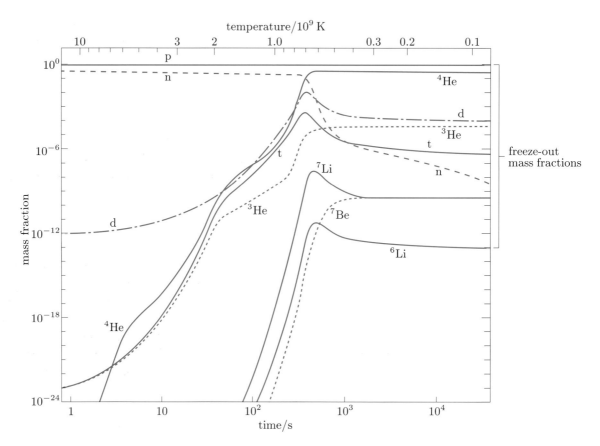

Figure 28 Evolution of nuclear abundances during the expansion of a typical model of the big bang. The vertical scale shows what fraction of the total mass the various types of nucleus form at different times. Note: H, He, Li and Be stand for the nuclei of hydrogen, helium, lithium and beryllium respectively, and the superscript indicates the mass of the isotope (e.g. ^{3}He is the isotope of helium containing two protons and one neutron. The deuteron is signified by d, and the nucleus of tritium (two neutrons + one proton) by t. (Smith, M. S., Kawano, L. H. and Malaney, R. A. (1993) *Astrophys. J. Suppl.* **85**, 219.)

10^9 K, but later (below 6×10^8 K) the concentration falls a little, because more deuterium is being used in making helium than is being synthesized from raw neutrons and protons. The point is that at this stage we are dealing with periods of time comparable to the mean lifetime of the neutron (930 s, i.e. about 15 minutes). Neutrons are being removed by decay — note the steady decline in its curve — and are thus no longer available for synthesizing deuterons. However, there are still collisions going on that are destroying the deuterons. From a cosmologist's point of view, this is the crucial stage. The *greater* the baryonic density of the Universe at that time, the longer the process of deuteron destruction can continue after the synthesis of deuterons has effectively ceased, and therefore, the *lower* the final concentration of deuterium. It is true that this effect is partly offset by the fact that in a denser Universe, more deuterium would have been formed in the first place. But the destruction of deuterium is more sensitive to density than is its initial formation.

Below about 4×10^8 K, all the nuclide mass fractions — apart from the decaying neutrons and tritons, the latter decaying to ^{3}He — are more or less constant because the thermonuclear reactions are then so slow. The most important feature of Figure 28 is the set of 'freeze-out' values of the elemental mass fractions to be found at the extreme right-hand side. This set of values corresponds to the particular assumption made about baryonic density. A different assumption for the baryonic density now (and hence at earlier times) would have led to a different set of curves, and these in turn would be expected to yield a different set of freeze-out values.

Objective 9   **SAQ 15**   Imagine a universe that differs from ours in that neutrons live twice as long, on average, as they do in our Universe, but the Hubble constant and the baryonic density have the same values as ours. On the basis of Reactions 15 to 20, would inhabitants of that universe expect to find more or less deuterium left over from the first 20 minutes of their universe than is left over from the first 20 minutes of ours?

Figure 29 shows how the computed freeze-out mass fractions of light nuclei depend on the assumed present-day value of the baryonic density. As you can see, the freeze-out abundance of ^{4}He is almost independent of the present baryonic density, whereas the freeze-out abundance of deuterium ^{2}H depends very sensitively on $\rho_{b,now}$. The most abundant nuclide of hydrogen (^{1}H) is not shown, but if it had been shown, it too would have been relatively insensitive to the present baryonic density, its mass fraction being about 0.75.

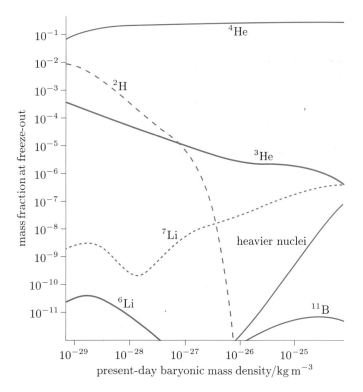

Figure 29   Predicted freeze-out mass fractions (at around 10^4 s after the big bang) for various light nuclei plotted against the assumed present-day mass density of baryonic matter, $\rho_{b,now}/c^2$.

7.3   Nuclear abundances as evidence for the big bang

What we have seen is that a theoretical model based on the assumption that there was a big bang, and incorporating an assumption about the present-day value of the baryonic density, $\rho_{b,now}$, leads to definite predictions as to what the nuclear abundances must have been when the elements froze-out. This, therefore, provides us with a third way of checking out the big bang hypothesis: Do the present-day cosmic nuclear abundances agree with these predictions for any plausible value of the present-day baryonic density?

Obtaining an answer to this question is not as easy as one might think. The trouble is that since the freeze-out abundances were established, about 20 minutes after the big bang, further modifications to the nuclear abundances have been going on. The story of most matter is that it exists for a few hundred million years as a rarefied gas, and then is slowly drawn into a star, where its nuclear composition is altered because it is heated up to temperatures at which further nuclear reactions take place. Because the temperature and density conditions in a star are very different from those encountered during the big bang epoch of *nuclear synthesis,* the thermonuclear reactions in stars are different, and they lead to a different mix of end products. Therefore, the freeze-out concentrations of the various elements *are not reflected directly in the abundances found in stars,* or indeed on the Earth which itself condensed out of stellar matter thrown out of stars during supernova explosions.

nuclear synthesis

The terms *nuclear synthesis* and *nucleosynthesis* can be used interchangeably.

Objective 9 **SAQ 16** Figure 30 shows in a schematic way the conditions of temperature and density under which cosmological nuclear reactions are important (region on the left), and the conditions for stellar nuclear reactions (region on the right).

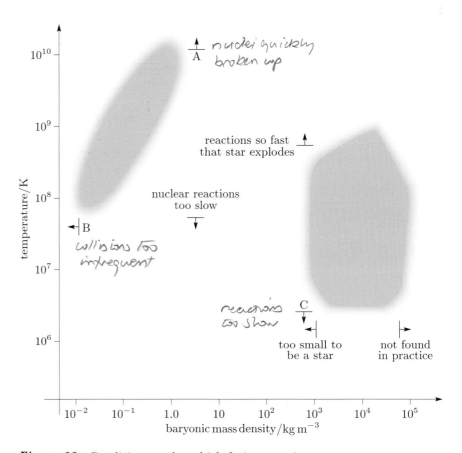

Figure 30   Conditions under which fusion reactions occur.

(a) Why is the area between A and B inclined to the axes rather than being vertical?

(b) Why does the area for the nucleosynthesis in the big bang not continue above A?

(c) Why does it not continue to the left of B?

(d) Why does the area for stars not continue below C?

The ways in which present-day observations of abundances can be compared with the predicted abundances at freeze-out are best considered by treating each of the relevant nuclei in turn.

First, we can note that there is good evidence from a wide range of astronomical bodies that the ^{1}H nuclide is the most common. This fits well with the freeze-out predictions, but is hardly conclusive. For more detailed insight we need to look at how the abundances of other nuclei (as expressed by their mass fractions) compare with that of ^{1}H.

Turning to the second most abundant nuclide, ^{4}He, we must confront the problem that much of the helium in the present-day Universe has been produced in stars. Fortunately, stars also produce other, heavier elements, and we can use these to determine how much of the helium in any region is the result of stellar processing rather than the big bang. Most stars are not hot enough to show helium lines in their spectra, and those that do are unsuitable for primordial abundance measurements. Instead, helium is best measured in the clouds of diffuse glowing gas that astronomers call HII regions, and even these clouds are best studied in smaller galaxies where there has been less stellar processing than in our own Galaxy.

One way of proceeding is to use spectral techniques to measure both the helium mass fraction and the relative abundance of oxygen nuclei for a number of extragalactic HII regions. Once the measurements have been made (and the uncertainties in those observations estimated) a plot such as that shown in Figure 31 can be compiled. By extrapolating the data back to zero relative abundance of oxygen, a 'primordial' value for the helium mass fraction can be deduced. Figure 31 implies a value of about 0.23 for this primordial helium mass fraction, in line with several other attempts to determine this quantity.

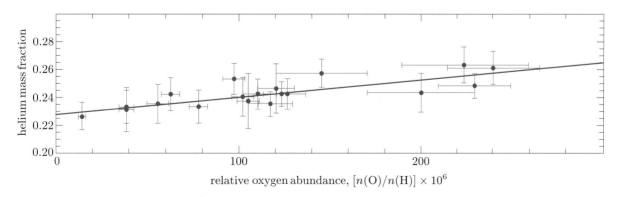

Figure 31   A plot of observed helium mass fraction against relative abundance of oxygen to hydrogen, expressed as $10^6 \times$ (number of oxygen nuclei)/(number of hydrogen nuclei), for several extragalactic HII regions. (Based on measurements by B. Pagel and colleagues.)

Comparing this 'observed' primordial helium mass fraction (0.23) with the predicted values in Figure 29, there is a reasonable level of agreement for a range of present-day baryonic densities.

The abundance of ^{3}He is less useful. A number of observational difficulties make it very hard to deduce anything reliable about the primordial abundance. The comparison of prediction and measurement has little to offer in this case.

The present-day abundance of $^7_3$Li can be deduced from spectral studies of metal-poor dwarf stars. The outer layers of these stars are believed to be relatively unchanged since they were formed. Most share the same lithium abundance despite having different amounts of other elements and different masses. This makes it probable that the lithium in these outer stellar layers has not been processed, and thus gives a direct indication of the primordial mass fraction of lithium. The observations favour a $^7_3$Li mass fraction close to 8×10^{-10}. According to Figure 29, this corresponds to a present-day mass density of baryonic matter of around $(1 \text{ to } 5) \times 10^{-28} \, \text{kg m}^{-3}$.

Finally consider the case of deuterium, $^2_1$H. Deuterium is destroyed in stars so its currently observed mass fraction provides a lower limit on its primordial mass fraction. This lower limit is about 3×10^{-5} to 7×10^{-5}, consistent (according to Figure 29) with present-day baryonic mass densities of about $5 \times 10^{-28} \, \text{kg m}^{-3}$ or less.

Bringing all of the observed abundances together, it does seem that they are consistent with the predicted primordial abundances (Figure 29), provided the present-day baryonic mass density is around $10^{-28} \, \text{kg m}^{-3}$.

Pleasingly, at least to those who like consistency, a present-day baryonic mass density of a few times $10^{-28} \, \text{kg m}^{-3}$ is in excellent agreement with the rather precise value of $\rho_{\text{b,now}}/c^2$ deduced by those who attempt to deduce cosmological parameters from the observed anisotropies in the cosmic microwave background radiation (as described in Section 6.3). The fact that there is a narrow range of values for the present-day baryonic densities in which the predicted and 'observed' light nuclear abundances agree, is a significant success for big bang cosmology. The fact that this narrow range of baryonic densities includes the value deduced by a quite different technique is a truly remarkable achievement.

8 Conclusion

We have seen that there are three independent pieces of evidence, all of which point to the occurrence of a big bang: (i) the recession of the galaxies; (ii) the cosmic microwave background; and (iii) the comparison between the calculated primordial nuclear abundances and the present-day composition of matter in the Universe. For these various reasons, the vast majority of cosmologists today accept that there was a big bang. This acceptance has been given extra support by the agreement between the value of the present-day baryonic mass density indicated by abundance measurements, and the value of that same quantity deduced from anisotropies in the cosmic background radiation.

In this Unit, we have seen two snapshots of the development of the Universe: the period of nuclear synthesis which shaped today's nuclear abundances; and the later period of decoupling from which emerged today's $3\,\mathrm{K}$ radiation. Later in this Block, you will see how these two epochs fit into the overall pattern of the Universe's evolution. One of the aims will be to try and push our description as far back in time as we can towards the very instant of the big bang; another will be to look as far ahead into the future as we can so as to uncover what might be the likely end of the Universe.

Summary

1 The visible matter in the Universe, stars and interstellar gas, is concentrated into galaxies, which are collections of $\sim 10^{11}$ stars. The galaxies themselves are usually to be found in clusters with typically tens or hundreds of members. The clusters, in their turn, are loosely associated in superclusters.

2 Distances are estimated via a series of intercalibrated techniques, each overlapping with and extending further than the previous one. The principal stages are:

> Earth–Moon and Earth–Sun distances involving radar ranging;
> Triangulation to nearer stars, using the diameter of the Earth's orbit as baseline;
> Calibration of luminosity against temperature for typical stars, using the stars of the Hyades cluster;
> Period–luminosity relationship for Cepheids, first for stars in our Galaxy, then in other galaxies;
> Type Ia supernovae, first in nearer, then in further-off galaxies;
> Classification of galaxies into recognizable types, of different luminosities.

Independent checks are provided by radio galaxies and observations of Type II supernovae.

3 The spectra of light emitted by the stars of distant galaxies is redshifted, such that the redshift, z, is proportional to the distance of the galaxy.

4 This cosmological redshift finds a natural explanation in terms of the galaxies receding from us (and from each other), in accordance with Hubble's law. This is the first indication that the Universe began with a big bang.

5 Radio astronomers have detected microwave radiation (i.e. radiation with wavelengths in the region of 1 cm) coming almost isotropically from all directions, with a thermal spectrum which tallies, as far as can be measured, with that expected from calculations based on the big bang model. No convincing alternative explanations have so far been advanced. This 3 K radiation therefore provides good evidence both for there having been a big bang, and for the isotropy of the Universe at the time when the radiation decoupled from matter, 4×10^5 years later.

6 If the Earth at present is moving with respect to the 3 K radiation, which we assume is the same as the 'rest frame' of the Universe at the time of decoupling, the angular distribution of the intensity of the radiation would be slightly distorted in a characteristic way. Measurements indicate that the Earth *does* have such a velocity, with a magnitude of about $400 \, \text{km s}^{-1}$.

7 If galaxies, or clusters of galaxies, had already begun to form at a time considerably less than 10^7 years after the big bang, one would have expected pre-existent inhomogeneities in the matter distribution to have left a trace of their presence by imparting some intensity variation to the angular distribution of the 3 K radiation. Such inhomogeneities have now been detected at the level of 1 part in 100 000.

8 If there has been a hot big bang, we would expect the material produced to consist of 23% by mass of helium, the rest being mostly hydrogen, with traces of other light elements.

9 Detailed comparisons of predicted primordial nuclear mass fractions with those deduced from observations also constrain the value of the present-day baryonic mass density. A value of 4 or $5 \times 10^{-28} \, \text{kg m}^{-3}$ fits well with both the nuclear abundance data and the data on anisotropies in the cosmic microwave background radiation.

Acknowledgements

Grateful acknowledgement is made to the following sources for material used in this Unit:

Figures 2, 3, 6, 9, 11 and 12b: from Hale Observatories; *Figure 4a,b*: based on H. Arp (1960) in *Astronomical J.*, **65**, 426; *Figure 5*: from *Le Grand Atlas de l'Astronomie*, Encyclopaedia Universalis editeur; *Figure 8*: based on Erik Holmberg in E. M. and G. R. Burbidge (1975) *Galaxies and the Universe*, University of Chicago Press; *Figure 10*: NASA/IPAC Extragalactic Database (NED), Jet Propulsion Laboratory, California Institute of Technology; *Figure 12a*: courtesy of Uppsala University; *Figure 13*: based on E. P. Hubble (1929) in *Proc. Nat. Acad. Sci.*, **15**, 168; *Figure 15*: adapted from A. Sandage (1970) *Physics Today*, vol. 34, February 1970, American Institute of Physics; *Figure 17a*: based on D. P. Woody *et al.* (1975) in *Phys. Rev. Lett.*, **34**, 1036–9; *Figures 17b and 25*: based on R. A. Muller (1978) in *Scientific American*, **238**, 64–74; *Figure 18*: courtesy of NASA/CSFC and COBE Science Working Group; *Figure 22*: courtesy of Dr A. A. Penzias; *Figure 26*: courtesy of NASA Goddard Space Flight Center, Greenbelt, Maryland; *Figure 27*: based on C. L. Bennett *et al.*, (2003) *The Astrophys. J. Supp.*, **148**(1), 1–27; *Figure 28*: from M. S. Smith, L. H. Kawano and R. A. Malaney (1993) *The Astrophys. J. Supp.*, **85**, 219–47; *Figure 29*: adapted from R. V. Wagoner (1973) *The Astrophys. J.*, **179**(2), 349; *Figure 31*: B. E. J. Pagel *et al.* (1992) *Monthly Notices of the RAS*, **255**, 325–45.

Self-assessment questions — answers and comments

SAQ 1 **(a)** At position A (Figure 32), the spectrum of the observed light consists of a flat continuum with absorption lines (corresponding to the transitions between the ground state and other states), as in Figure 33(a).

(b) At position B, no direct light is seen. The light that is observed comes from calcium atoms that have been excited, and is therefore an emission spectrum of lines on a dark background, with the wavelengths of the lines the same as those of the absorption lines seen at position A. This is shown in Figure 33(b).

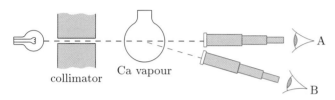

Figure 32 White light is passed through a bulb containing calcium vapour and observed at A (in the direction of the beam) and at B (at an angle to the beam).

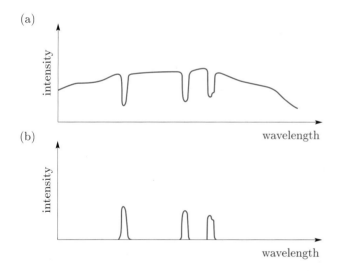

Figure 33 The spectra observed at the points A and B of Figure 32.

SAQ 2 **(a)** The result $\Delta f/f = -gH/c^2$, given in Unit 9, applies only to small changes in height, H. For a large change, one must take account of the fact that g changes with height. The value of g is given by equating the weight of a body, mg, to the gravitational attraction. Thus at the surface of the Earth

$$mg = \frac{GMm}{R^2} \quad \text{so that} \quad g = \frac{GM}{R^2}$$

where M and R are the mass and radius of the Earth respectively. At any distance r (greater than R), we may write more generally

$$g(r) = \frac{GM}{r^2}.$$

At each distance r, the frequency shift caused by rising to $r + dr$ is given by the formula of Unit 9, (p 38, 39)

$$\frac{\mathrm{d}f}{f} = -\frac{GM}{r^2 c^2}\,\mathrm{d}r$$

where we write dr in place of H. Thus the total frequency shift suffered in escaping to a very large distance ('infinity') is determined by:

$$\int_{f_1}^{f_0} \frac{\mathrm{d}f}{f} = \int_{R}^{\infty} -\frac{GM}{r^2 c^2}\,\mathrm{d}r$$

where f_1 is the emitted frequency and f_0 is the observed frequency. This gives the method asked for in the SAQ.

Performing the integration (which the SAQ did *not* ask for) gives the stated result. In fact,

$$\int_{f_1}^{f_0} \frac{\mathrm{d}f}{f} = \left(+\frac{GM}{rc^2} \right)_{R}^{\infty} = -\frac{GM}{Rc^2}.$$

Provided the magnitude of $\Delta f = f_0 - f_1 \ll f_1$, we can approximate the left-hand side of this equation by

$$\int_{f_1}^{f_0} \frac{\mathrm{d}f}{f} = \int_{f_1}^{f_1+\Delta f} \frac{\mathrm{d}f}{f} \approx \frac{\Delta f}{f_1}$$

and conclude that, for any emitted frequency f,

$$\frac{\Delta f}{f} = -\frac{GM}{Rc^2}.$$

(b) **(i)** Using the result of part (a), we can find the effect of the gravitational redshift on light leaving the star as follows:

$$\frac{\Delta f}{f} = -\frac{GM}{Rc^2} = -\frac{6.67 \times 10^{-11} \times 2 \times 10^{30}}{7 \times 10^8 \times (3 \times 10^8)^2}$$

$$= -2.12 \times 10^{-6}.$$

All the radiation leaving the star has been 'redshifted' by this amount by the time it reaches the Earth, which we may certainly regard as at an 'infinite' distance from the star.

(ii) The speed at the edge of the star is

$$v = \frac{2\pi R}{T} = \frac{2\pi \times 7 \times 10^8}{2 \times 10^6}\,\mathrm{m\,s}^{-1} = 2200\,\mathrm{m\,s}^{-1}.$$

As given in Unit 6, the Doppler shift is

$$\frac{f_0}{f_1} = \sqrt{\frac{1 + (v/c)}{1 - (v/c)}}$$

when v is a speed of approach. Here,

$$\frac{v}{c} = \frac{2200}{3 \times 10^8} = 7.33 \times 10^{-6}.$$

This could be used directly in the formula but since v/c is so small, we can use the approximation

$$\sqrt{\frac{1 + (v/c)}{1 - (v/c)}} \approx 1 + \frac{v}{c}$$

so the fractional change in frequency, $\Delta f / f$, is simply:

$$\frac{f_0 - f_1}{f_1} = \frac{f_0}{f_1} - 1 \approx 1 + \frac{v}{c} - 1 = \frac{v}{c} = 7.33 \times 10^{-6}.$$

Light from the advancing limb is blueshifted by this amount, and from the receding limb is redshifted by this amount. The width imparted to lines is thus the fraction 14.7×10^{-6} of their central frequency.

(iii) The turbulent motion again produces both blueshifts and redshifts, of amount $\pm 6000\,\mathrm{m\,s^{-1}}/c = \pm 20 \times 10^{-6}$.

The frequency of the 656 nm line is given by

$$f = \frac{c}{\lambda} = \frac{3 \times 10^8}{656 \times 10^{-9}}\,\mathrm{Hz} = 4.57 \times 10^{14}\,\mathrm{Hz}.$$

The frequency shifts are therefore as shown in Table 2.

TABLE 2

	$\Delta f / f$	$\Delta f / \mathrm{MHz}$
gravitation	-2.12×10^{-6}	-970
rotation	$\pm 7.33 \times 10^{-6}$	± 3350
turbulence	$\pm 20 \times 10^{-6}$	± 9150

SAQ 3 If λ_1 is the emitted wavelength and λ_0 is the observed wavelength, then

$$\frac{\Delta \lambda}{\lambda} = \frac{\lambda_0 - \lambda_1}{\lambda_1} = \frac{\lambda_0}{\lambda_1} - 1 = 0.02.$$

Therefore

$$\lambda_0 = 1.02 \lambda_1.$$

Thus the observed wavelengths are:

$1.02 \times 434 = 443\,\mathrm{nm}$;

$1.02 \times 486 = 496\,\mathrm{nm}$;

$1.02 \times 656 = 669\,\mathrm{nm}$.

The relative frequency shift is given by

$$\frac{\Delta f}{f} = \frac{f_0 - f_1}{f_1} = \frac{c/\lambda_0 - c/\lambda_1}{c/\lambda_1} = \frac{\lambda_1 - \lambda_0}{\lambda_0}.$$

But

$$z = \frac{\lambda_0 - \lambda_1}{\lambda_1}.$$

Therefore

$$\lambda_0 = (1 + z)\lambda_1$$

and

$$\frac{\Delta f}{f} = \frac{\lambda_1 - (1+z)\lambda_1}{(1+z)\lambda_1} = -\frac{z}{1+z}$$

$$= -\frac{0.02}{1.02} = -0.0196.$$

Note that the shift in frequency is the same for all lines of the spectrum, but differs from the shift in wavelength, which is z.

SAQ 4 At 2×10^6 ly, the fraction of light intercepted is: $\dfrac{\text{area of telescope}}{\text{area of sphere of radius } 2 \times 10^6\,\text{ly}}$

$$= \frac{\pi \times 1^2}{4\pi(2 \times 10^6 \times 9.46 \times 10^{15})^2} = 6.98 \times 10^{-46}.$$

The energy received from the faintest star is 3.2×10^{-17} W.

Therefore the output of the star must be

$$\frac{3.2 \times 10^{-17}}{6.98 \times 10^{-46}}\,\mathrm{W} = 4.6 \times 10^{28}\,\mathrm{W}.$$

From Figure 4(b), the corresponding period of the Cepheid variable is about 3 days.

SAQ 5 Both statements are incorrect.

(i) The overall luminosity of a cluster cannot be used as a standard lamp because, as stated in Section 3.2, clusters can vary enormously in size from thousands of galaxies to a single galaxy. All one can say is that the luminosity of a *typical* galaxy in that cluster (averaged over a number of the galaxies it contains) is the same for each cluster.

(ii) The shell thrown out from the supernova cannot be resolved optically. Instead, the radius of the fireball at a given moment is *calculated* from the estimated speed of the ejected material and the time that has elapsed since the explosion. The colour (temperature) determines the light output per unit area. Hence, knowing the surface area from the radius, one arrives at an estimate of the luminosity of the supernova at the given time after the explosion. This luminosity is then compared to the measured flux density to obtain the distance to the star.

SAQ 6 To get a redshift that can confidently be regarded as due to expansion, we must look to another cluster.

Intercluster distances are approximately

$$2 \times 10^8\,\text{ly} \approx (2 \times 10^8)(10^{16}\,\text{m}) = 2 \times 10^{24}\,\text{m}.$$

The Hubble constant is about $2 \times 10^{-18}\,\mathrm{s^{-1}}$.

Hence the cosmological redshift of a neighbouring cluster is

$$z = \frac{Hr}{c} \approx \frac{(2 \times 10^{-18})(2 \times 10^{24})}{3 \times 10^8} \approx 0.01.$$

Compare this with Figure 13 — Hubble was not able to look far enough!

SAQ 7 From SAQ 2(b), parts (ii) and (iii), the minimum width of the lines is

$$\Delta f \approx (\pm 20 \pm 7) \times 10^{-6} f$$

with a systematic fractional shift of about -2×10^{-6} (SAQ 2(b), (i)). Thus the contributions to $\Delta f / f$ due to turbulence, rotation and gravitational redshift vary between $(+20 + 7 - 2) \times 10^{-6}$ and $(-20 - 7 - 2) \times 10^{-6}$. This gives an error of about $\pm 3 \times 10^{-5}$. For a galaxy at 4×10^6 ly $(\approx 4 \times 10^6 \times 10^{16}\,\text{m})$, the expected cosmological redshift is

$$z = \frac{Hr}{c} \approx \frac{(2 \times 10^{-18})(4 \times 10^6 \times 10^{16})}{3 \times 10^8}$$

$$\approx 3 \times 10^{-4}.$$

Thus the line-broadening effects are approximately a fraction

$$\frac{3 \times 10^{-5}}{3 \times 10^{-4}} = 0.1$$

of the expected cosmological shift.

Notice that a distance of 4×10^6 ly corresponds to nearby galaxies in our Local Group and the relative uncertainties in redshift will *decrease* with distance. They are commonly less than 1%. The uncertainties in distance, however, will increase with distance, because less accurate estimation procedures have to be used. Also note that a distance of 4×10^6 ly may be unreasonably small for cosmological consideration.

SAQ 8 One can use any pair of points on the straight line graph that has been drawn, but to minimize reading errors we take the extreme points. The difference in z is $1.0000 - 0.0025 = 0.9975$. The difference in distance, r, is $4000 - 10 = 3990$ Mpc. Thus,

$$\frac{z}{r} = \frac{0.9975}{3990} = 2.5 \times 10^{-4}\,\text{Mpc}^{-1}.$$

But

$$z = \frac{Hr}{c}, \qquad \text{i.e.} \qquad H = \frac{zc}{r}.$$

So

$$H = 2.5 \times 10^{-4}\,\text{Mpc}^{-1} \times 3 \times 10^5\,\text{km s}^{-1}$$
$$= 75\,\text{km s}^{-1}\text{Mpc}^{-1}.$$

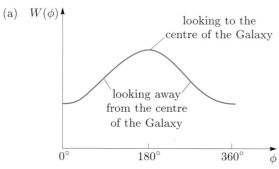

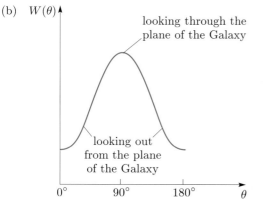

Figure 34 Variation of intensity of radio waves with angle (a) in the plane of the Galaxy, and (b) in a plane perpendicular to the Galaxy.

SAQ 9 **(a)** If ϕ is the angle between the direction of the telescope and the line from the centre of the Galaxy to the telescope, with $\phi = 0$ corresponding to looking directly away from the centre of the Galaxy, then, if the telescope is swept round in the plane of the Galaxy, the variation of intensity W of radio waves with ϕ will be that shown in Figure 34(a).

(b) If θ is the angle between the direction of the telescope and a line drawn perpendicular to the plane of the Galaxy at the telescope, with $\theta = 0$ corresponding to looking directly 'upwards' through the disc, then the variation of intensity W of radio waves with θ will be as shown in Figure 34(b).

SAQ 10 **(a)** Let $y = W/T^3$ and $x = f/T$. Then

$$y = \frac{A(f/T)^3}{\exp(Bf/T) - 1} = \frac{Ax^3}{\exp(Bx) - 1}$$

which shows that W/T^3 is a function of f/T only.

(b) From the above expression for y, we see there must be a universal value of x which gives a maximum value of y. But a maximum value of y implies a maximum value of W at a given temperature. If we denote this value of x by the constant $x_{\max}$, it follows that

$$x_{\max} = \frac{f_{\max}}{T}$$

and hence that

$$f_{\max} = x_{\max}T \propto T.$$

Thus the frequency, $f_{\max}$ which gives maximum intensity is proportional to the temperature T. (You might have heard of this relation referred to as Wien's displacement law.)

(c) The explanation is that the cosmic background radiation has cooled during the time it took for the radio message to reach you. The discrepancies between their results and yours can all be explained by this difference in temperature of the radiation.

In part (b), it was shown that the frequency, $f_{\max}$, corresponding to maximum intensity, is proportional to T. Since their value of $f_{\max}$ is twice yours, the temperature of the cosmic background radiation then must have been twice what it is here and now, i.e. it must have been just under 6 K.

In part (a), it was shown that:

$$y = \frac{W}{T^3} = \frac{Ax^3}{\exp(Bx) - 1}.$$

If the maximum intensity is $W_{\max}$, it follows that

$$\frac{W_{\max}}{T^3} = \frac{Ax_{\max}^3}{\exp(Bx_{\max}) - 1}$$

which is independent of T since $x_{\max}$ is a universal constant from part (b). Hence

$$W_{\max} \propto T^3$$

which is why their value of $W_{\max}$ is $2^3 = 8$ times yours.

Finally, Equation 8 shows that the intensity at a given low frequency is proportional to the temperature and hence will be twice what you measure now at the *same* low frequency.

SAQ 11 (a) We have already established (in part (b) of the previous question) that the frequency of maximum intensity increases proportionately with temperature. As yellow light has a higher frequency than red light we would expect the yellow star to be the hotter.

(b) From Figure 19, we see that the maximum intensity for a temperature of, say, 1646 K occurs at a frequency of about 92×10^{12} Hz. The maximum for room temperature, taken to be 300 K, is then given by $(300/1646) \times 92 \times 10^{12}$ Hz $= 16.8 \times 10^{12}$ Hz.

Radiation of this frequency occurs in the infrared part of the spectrum.

SAQ 12 $\dfrac{\lambda_0}{\lambda_d} \approx z = 1000 = \left(\dfrac{t_0}{t_d}\right)^{2/3}$.

Therefore,

$$t_d = \frac{t_0}{1000^{3/2}} = 4.43 \times 10^5 \text{ years.}$$

SAQ 13 Proxima Centauri is the nearest star, and so is undoubtedly in our Galaxy. Our Galaxy is rotating, as is suggested independently of all frames of reference by its disc-like shape (recall also the Newton bucket experiment described in the Appendix to Unit 2). So the line to Proxima Centauri is *not* a good x^1-axis, because of the relative motion of the Sun and Proxima Centauri within the Galaxy.

The Andromeda galaxy is the nearest galaxy to ours (apart from satellite galaxies and the Magellanic Clouds). It is gravitationally bound to our Galaxy and moving relative to us. The period of the mutual 'orbit' of our Galaxy and the Andromeda galaxy is certainly longer than the period of the Sun (or Proxima Centauri) round our Galaxy, because the distance from our Galaxy to the Andromeda galaxy is about a hundred times bigger than the distance of the Sun (or Proxima Centauri) from the centre of our Galaxy. Nevertheless, the line connecting our Galaxy to the Andromeda galaxy cannot be taken as fixed because of their motions within the Local Group.

Thus methods (i) and (ii) are not equally acceptable; (ii) is preferable to (i), but is still not fully satisfactory.

SAQ 14 The mean speed of the Earth round the Sun can be obtained by taking the orbit as a circle of radius 1.5×10^{11} m. The speed v is then

$$v = \frac{2\pi \times 1.5 \times 10^{11}}{365.25 \times 24 \times 60 \times 60} \text{ m s}^{-1}$$
$$\approx 30 \text{ km s}^{-1}.$$

(Strictly, the orbit is not quite circular, so the speed varies somewhat above and below this value.)

From Equation 12 in Section 6.2, the maximum change in apparent temperature, ΔT, which this motion can cause at low frequencies arises when $\cos\theta = 1$. $\Delta T/T$ is then a fraction

$$\frac{v}{c} \approx \frac{3 \times 10^4}{3 \times 10^8} = 10^{-4}.$$

The speed of the Solar System round the Galaxy, about 230 km s^{-1}, gives a $\Delta T/T$ of about $(2.3 \times 10^5)/(3 \times 10^8) \approx 10^{-3}$.

SAQ 15 Other things being equal, the increase in neutron lifetime will mean that more deuterium is formed in the reaction

$$n + p \longrightarrow d + \gamma \qquad \text{(Eq. 15)}$$

and more is destroyed in the reaction

$$n + d \longrightarrow {}^{3}\text{H} + \gamma \qquad \text{(Eq. 19)}$$

eventually producing ^{4}He in the reaction

$$p + {}^{3}\text{H} \longrightarrow {}^{4}\text{He} + \gamma. \qquad \text{(Eq. 20)}$$

In their universe, the increase of neutron lifetime will mean that at any point in time there will be more neutrons in their universe than in ours. Moreover, the *ratio* of neutrons in their universe to neutrons in ours will steadily increase with time. Thus although at the time when Reaction 15 is important their rate of formation of deuterium will be somewhat faster than ours, by the time the later Reaction 19 comes into its own their rate of destroying deuterium will be greater still in comparison with ours. Thus, deuterium will be formed more readily, but *destroyed even more readily*. The net result will be that they find *less* deuterium than we do.

SAQ 16 (a) While the Universe was expanding and the density decreasing during the big bang, the temperature dropped. (A universe that was perfectly uniform at all stages would be shown as a line between A and B. We have chosen to show a narrow area instead, corresponding to some non-uniformities of density and temperature.)

(b) Above A, collision energies were too high for nuclei to hold together.

(c) Because of the reduced density to the left of B, collisions occur too infrequently to have much importance.

(d) Below C, because the nuclei are moving slowly, they have difficulty in overcoming the electrostatic repulsion between their charges, and thus getting close enough for fusion to take place. For this reason, fusion reactions proceed only slowly and the energy release is not sufficient to make the body glow (and by definition, stars give off light). This region would correspond to planets, rather than stars. Jupiter, for instance, our largest planet, misses being a star by a factor of about 50.

These additional conditions have been included in Figure 35 (overleaf).

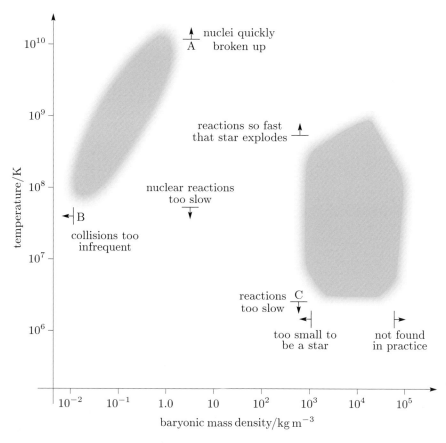

Figure 35 Further limits on fusion reactions.

Unit 14 General relativity and cosmology

Prepared by the Course Team

Contents

Aim

Show how the structure of spacetime can be modelled by a metric based on a homogeneous Universe.

Objectives

When you have finished studying this Unit, you should be able to:

1 Describe the evidence that points to the homogeneity of mass in the Universe, at least over the distance scale of relevance to cosmological models. Explain what is meant by the cosmological principle.

2 Given the Robertson–Walker metric, explain the meaning of the various symbols appearing in it.

3 Distinguish between curved space and curved spacetime.

4 Compare the Robertson–Walker metric with (i) the Schwarzschild metric, and (ii) the flat metric of special relativity.

5 Using the Robertson–Walker metric, derive a relationship between the cosmological redshift, z, and the scale factor, $R(t)$.

6 Explain the meaning of the deceleration parameter, q, and recognize definitions of q and H expressed in terms of $R(t)$.

Study comment

You should note that the work load associated with this Unit is about two-thirds that of an average Unit. You are advised, therefore, to make an early start on Unit 15 which will contain about four-thirds of a Unit's worth of material.

---

1   Introduction

Band 3 of AC4 introduces this Unit.

In this Unit, we start to apply the general theory of relativity to the Universe as a whole. The type of metric we eventually come up with will depend on what we assume the mass distribution of the Universe to be. From the outset, it should be realized that there has always been the hope that the distribution will turn out to be very nearly homogeneous; if it is not so, the treatment becomes very much more difficult and complicated.

We already have one indication (from the variations in the cosmic microwave background radiation) that the distribution is not entirely isotropic. However, the lack of isotropy was found to be exceedingly small; it was consistent with having come from a highly isotropic distribution of matter.

But it is not sufficient to know (or assume) that the matter is distributed isotropically. This alone does not define the metric. To see why, it may be useful to pause a moment and make a comparison with the other metric treated so far, the Schwarzschild metric as introduced in Unit 10 & 11. That metric is for the spacetime outside a spherically symmetrical body (i.e. a body which is isotropic about its centre), and was extensively used in the last Block. Only the total mass of the body entered into the metric — the radial distribution of density was irrelevant as far as the metric *outside* the body was concerned, and the metric would not apply inside.

But now we are looking for the metric *inside* a body — inside the Universe. To do this, it is necessary to know the radial distribution as well as angular distribution of density. Thus, our first task is to investigate directly how the galaxies are distributed throughout space.

2 The distribution of clusters of galaxies

We have already mentioned, in Section 3.2 of Unit 13, that galaxies occur in clusters, our own Milky Way galaxy being a member of the Local Group. This cluster consists of over 30 galaxies occupying a volume of space about 2 Mpc, or 2 million parsecs across (a parsec, or pc, being about 3 light-years). For comparison, the diameter of the disc of the Milky Way is 0.03 Mpc. Although the galaxies are moving relative to one another, they are gravitationally bound together, the grouping thus being a more-or-less permanent feature.

Beyond the Local Group, there are other clusters of galaxies. One of the nearest is to be found at a distance of 3 Mpc. Figure 1 shows our near neighbours.

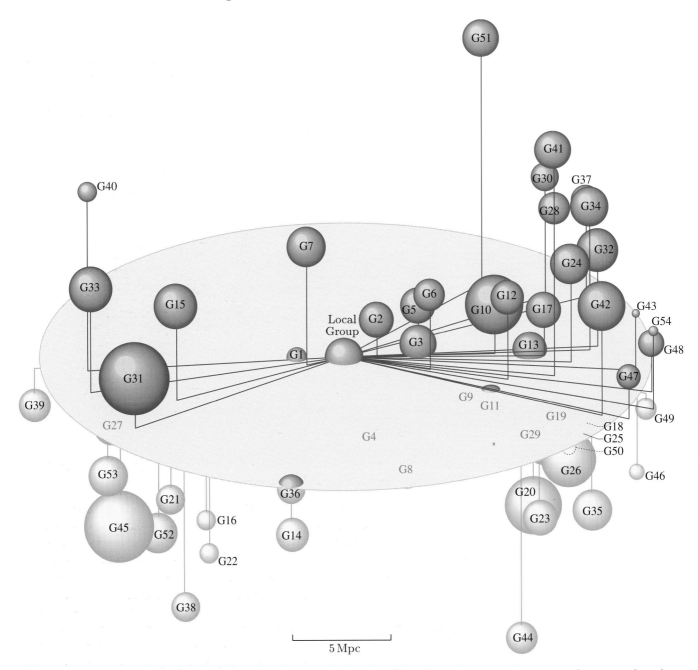

5 Mpc

Figure 1   The spatial distribution of nearby clusters of galaxies. (The G numbers are cluster catalogue numbers.)

Clusters are of widely different sizes, ranging from small ones like our Local Group, to rich clusters consisting of several thousand galaxies. Broadly speaking the volume occupied by a cluster is independent of the number of galaxies it contains, the more richly populated ones having their galaxies more tightly packed together.

Objective 1   SAQ 1   Sketch the broad features of the sky as it would be seen from a point in intergalactic space within a cluster of galaxies.

Clearly, if we look at the distribution of matter on the scale of a galaxy, or of a cluster, or even on the scale shown in Figure 1, it cannot be regarded as uniform. If we were to continue to go up in scale and were to take in ever larger volumes of space, do we come to a scale where we can say that at least the clusters are more or less evenly distributed throughout space? The answer is no. Since the mid-1970s it has been known that the clusters are themselves loosely associated into so-called superclusters. Our Local Group belongs to the fairly typical supercluster shown in Figure 2. It is 25 to 50 Mpc across and consists of about 1000 bright galaxies and a much larger number of fainter ones.

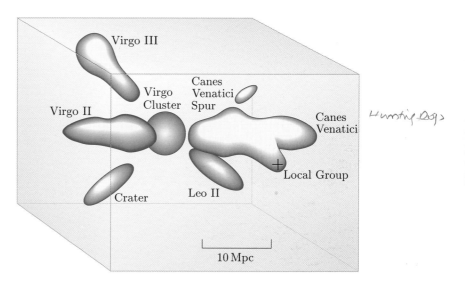

Hunting Dogs

Figure 2   The Local Supercluster

Figure 3 shows the result of a survey of about two million bright galaxies covering about 10% of the celestial sphere.

The bright patches show the superclusters as often elongated regions surrounding huge volumes that are almost devoid of galaxies. Although this is just a two-dimensional picture, the inhomogeneities are clearly seen. But one can do better. Using the redshift of the galaxy as a measure of its distance from us, we can investigate inhomogeneities along the line of sight. Several surveys have built up a three-dimensional picture of the distribution of galaxies. These confirm that the clusters are gathered into superclusters that are often long and winding, with others forming vast curved sheets and walls enclosing voids. One particularly notable feature is called the Great Wall. It covers an area of 200 Mpc by 80 Mpc and is probably no more than 6 Mpc thick.

Figure 3 A survey of the positions of two million galaxies carried out by a team of astronomers from Oxford University (1990).

Figure 4 A picture taken by the Hubble Space Telescope of a tiny part of the sky no bigger than 1/60th of the diameter of the Moon. This is so narrow that only a few of the images are those of foreground stars belonging to our Milky Way Galaxy. Most of the objects are distant galaxies, the faintest being close to the visible horizon of the Universe.

Figure 4 shows an example of a deep-field picture taken in 1996 by the Hubble Space Telescope. Other 'deep fields' have been recorded since. All show very distant galaxies in large numbers.

Another type of survey is that in which one probes deeply into space over a narrow region of the sky; these are called 'borehole' surveys. The results of distance measurements to superclusters along one of these 'boreholes' is shown in Figure 5.

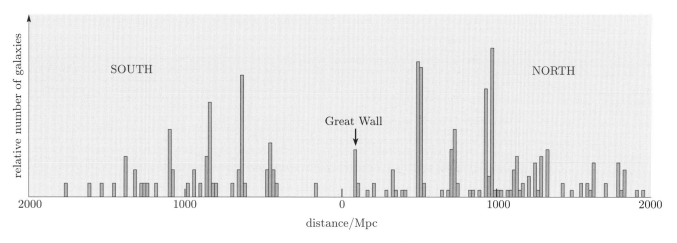

Figure 5   A 'borehole' survey locating the positions of superclusters along the north and south directions from the Milky Way galactic plane.

The measurements extend 2000 Mpc towards the galactic north and south poles (where obscuration from matter within our Milky Way Galaxy is at a minimum). The Great Wall shows up as a very clear peak. The existence of other peaks is taken as an indication that the probing survey is passing through more distant superclusters, some of them possibly also being in the form of a wall.

In summary, the superclusters taken together resemble a gigantic sponge, with the superclusters themselves making up the substance of the sponge while the voids between superclusters play the role of the sponge's pores. Thus, on a scale up to several hundred Mpc the distribution of matter in the Universe is rather inhomogeneous, consisting of walls, and filaments surrounding voids.

Nevertheless — and this is the crucial point — several hundred Mpc is still small compared to the overall dimensions of the observable universe (5000 Mpc). At such distances, the features of the Universe do seem to be broadly similar in whichever direction one looks. As we are interested in developing a metric to describe behaviour on the largest of all scales, it is concluded that for our particular purposes, it is indeed reasonable to assume an isotropic distribution. Moreover, from the borehole surveys, we begin to have something of an indication of homogeneity in directions pointed radially away from us. This latter point, however, we examine more closely in the next section.

3 Radial distribution and the metric

metric

We are considering the broad overall structure of the Universe because it will influence the *metric* for its spacetime. If the number of galaxies per unit volume is the same everywhere, i.e. the Universe is homogeneous, then the radial number density function $\rho(r)$, is constant. But so far we have not actually specified how we define the distance r, and how we find the element of volume with sides Δr, $r\,\Delta\theta$, and $r\sin\theta\,\Delta\phi$. (Here, θ is the latitude and ϕ the azimuthal angle.) In Euclidean space, there is no problem — the area of a sphere of radius r is $4\pi r^2$, so the total number of galaxies within a shell of thickness Δr is $\rho(r)4\pi r^2\Delta r$. But if space is curved, then the volume element need not be given by the formula for *Euclidean* geometry. This introduces an extra factor into the expression for the mass of galaxies per unit volume. If the metric is known, this factor can be calculated. But this implies you need to *know* the metric before you can analyse the data to *find* the metric!

In practice, the analysis consists of taking various possible models for space, calculating for each of them what values of observed density they would give as a function of r, and comparing these calculations, one by one, with the data. Let's follow that method through for the most straightforward case: Euclidean space and a homogenous universe. Since the number of galaxies in a thin spherical shell is $\rho 4\pi r^2\Delta r$, the number of galaxies at a distance less than, say, a is $\rho\frac{4}{3}\pi a^3$ — the well-known expression for the volume of a sphere multiplied by the constant density. Now, if all galaxies were of the same luminosity L, then the observed brightness of each galaxy at the Earth (its flux density, S) would be proportional to L/r^2, where r is the distance from the galaxy to the Earth. This leads to the result that the number of galaxies with a flux density greater than a given value, say S, is, in this model, proportional to $S^{-3/2}$.

Objective 1　　　　**SAQ 2**　Derive the $S^{-3/2}$ law, for a homogenous distribution of identical galaxies.

The $S^{-3/2}$ law can be plotted to give the straight line labelled 'Euclidean' in Figure 6, and compared with experimental data. Already by 1968, when these observations were made with the Cambridge radio telescope, it was clear the data were not in agreement with the $S^{-3/2}$ law.

What is the reason for this failure? Several possibilities deserve to be considered. First, the distance ladder involving radio galaxies rests on the assumption that all radio galaxies are identical; i.e. we have assumed that all galaxies have the same luminosity, L. This is clearly not true. Even more serious are the systematic effects arising from the finite speed of light. When we look at distant galaxies we are seeing them as they were a long time ago. This is because their light takes time to reach us. The time delay is, of course, numerically equal to the distance in light-years. The farther away the galaxy, the younger it is likely to be, as seen by us. Does this mean that the farthest galaxies should be assigned greater or lesser luminosities than the nearer ones? The answer to this question depends on our theories of the evolution of galaxies, and, as we explained in Section 2 of Unit 13, this is an unresolved problem.

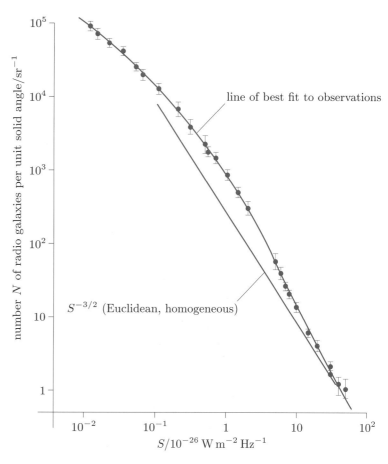

Figure 6   The number, N, of radio sources per steradian (sr) with received power greater than S. One steradian (1 sr) is the solid angle subtended at the centre of a sphere of unit radius by a portion of the surface having unit area.

Another possible reason for the failure of the $S^{-3/2}$ law is that space is not Euclidean. Indeed, in view of the relationship, in general relativity, between the curvature of spacetime and the distribution of mass, we might anticipate that space might well *not* be Euclidean, and as a consequence, the total number of galaxies within the volume of a spherical shell of thickness Δr is *not* given by $4\pi\rho r^2 \Delta r$.

The third possibility is that the gross distribution of matter in our Universe is *not* homogeneous. If this were true it would lead to great complexities, so cosmologists have resisted settling for this option.

We can sum up our discussion of the distribution of matter by saying that, whereas it is relatively easy to check out its homogeneity in different directions *across* the sky, there are considerable difficulties checking it out in terms of distances measured *along* any line of sight. The experimental results are certainly consistent with homogeneity, but hardly convincing.

In view of these difficulties, we adopt a different strategy. We ask the question: Why *should* the density of mass vary as a function of its radial distance from us? Such a dependence would put us here on Earth in a privileged position in the cosmos; we would be at the centre of the Universe. There is, of course, no reason for believing any such thing. So, without further ado, we make the assumption that the Earth occupies no special position, and that the Universe looks essentially the same from any chosen vantage point. In other words, we make an appeal to a *symmetry principle*. You recall from Block 1 how physicists use the term 'symmetry' to describe a situation where on making a certain transformation (e.g. a translation or rotation in space, or a linear boost) the laws of physics remain unaltered. Now we are extending that idea to say that if one were to change one's position in space, the Universe would look essentially the same — its mass would continue to be homogeneously and isotropically distributed. We call this symmetry principle the *cosmological principle*.

cosmological principle

> The cosmological principle is the hypothesis that the Universe is at all times homogeneous and isotropic.

The principal evidence in favour of the cosmological principle is to be found in the extreme isotropy of the cosmic microwave background radiation, with strong supporting evidence from the observed isotropy of clusters of galaxies. If the Universe is isotropic from all vantage points, it must also be homogeneous.

The remainder of this Block is based on the assumption that the cosmological principle is valid, and that the laws of physics observed on Earth apply throughout the whole Universe at all times.

4 Analysis of a uniform isotropic expanding universe

The next task is to find the metric that describes the geometry of the spacetime generated by an isotropic and homogeneous distribution of matter — moreover, a distribution of matter that is expanding according to Hubble's law.

The metric for an isotropic homogeneous spacetime was derived independently by H. P. Robertson, and by A. G. Walker in 1935/36. Since that time, the Robertson–Walker metric has become a cornerstone of cosmological theory.

Our treatment of the Robertson–Walker metric will be on the same footing as our analysis of the Schwarzschild metric in Unit 10 & 11. We shall approach it by the following three stages:

1 A description of the coordinate system used by Robertson and Walker to label events (Section 4.1).

2 A quotation of the form of the Robertson–Walker metric in this coordinate system (Section 4.2).

3 A discussion of the geometric properties of spacetime implied by this metric and coordinate system (Section 4.4).

4.1   Robertson and Walker's comoving coordinates

As explained in Block 3, the choice of coordinates is a crucial step in the formulation of problems in general relativity. An inconvenient choice could lead to severe mathematical difficulties.

In the case of the Schwarzschild metric for the structure of spacetime around a spherically symmetric star, we were led by the symmetry of the problem to the use of spherical polar coordinates (t, r, θ, ϕ). The time coordinate, t, was defined with respect to a standard clock placed beyond the range of the star's gravitational influence.

In the case of the large-scale structure of spacetime in our Universe, we have three points to guide us:

1 The distribution of matter is assumed to be isotropic from our vantage point on Earth.

2 The distribution of matter is assumed to be homogeneous; one can no longer have a standard clock placed far from all gravitational influence.

3 Each cluster of galaxies is assumed to behave like a point particle in free fall in the underlying spacetime created by all the others.

These points led Robertson and Walker to propose the following coordinate system:

Strictly speaking, we should subtract the motions of the Earth round the Sun, the Solar System round the Galaxy, and our Galaxy within its cluster, because we are not interested in such local motions.

The time and position of a cluster of galaxies, relative to a telescope on Earth, are denoted by $(t, \sigma, \theta, \phi)$. θ and ϕ are the usual angular variables of a spherical polar coordinate system. The galaxy clusters move radially outwards from the Earth (as they do from *any* vantage point) so these angles do not change as the Universe evolves. The time coordinate, t, in the Robertson–Walker coordinate system is defined by assigning *identical synchronized* clocks to each galaxy cluster. The time coordinate in the region of spacetime around a given clock is taken to be the reading, t, recorded by that clock as it falls freely along the world-line of its cluster. The problem of synchronizing all the clocks so that, in a certain sense, they all run from the same zero of time, is not a trivial one. But on the enormous time-scales with which we are concerned in cosmology, it is sufficient to choose some ubiquitous occurrence — such as the stage of the Universe when electrons and protons first came together to form hydrogen atoms — to mark the zero of time everywhere. In any case, two Robertson–Walker observers in different clusters can always *check* whether their clocks are synchronized if each of them agrees to send the other a light signal when their own clock reads, say, 2×10^9 years exactly. In general, if the clocks are *not* synchronized, the two observers will *disagree* about the times (on their clocks) when they receive one another's signals.

The most difficult coordinate to comprehend is σ:

It must be understood that σ is **NOT** *the radial distance from the Earth to the cluster. Instead, σ is taken, by definition, to be constant throughout all time for any given cluster.*

comoving coordinate   Coordinates such as σ are known as *comoving coordinates*, since they 'move along' with the particles whose positions they label. Of course, the constancy of σ for a cluster does *not* mean that the cluster remains at a constant distance from Earth. The mere choice of coordinate system tells us nothing about distances — which can only be defined after a metric has been specified. In order to clarify this point, we shall pause to consider some rather more familiar examples of comoving coordinates in two spatial dimensions.

Imagine a chessboard marked on a flat rubber sheet, with a pawn at, say, one edge of the board. If the sheet were stretched (uniformly, by four bars along the edges) the pawn would remain at one edge. In fact the positions of the pieces as given by their coordinate labels would be unaffected, and the game would not be changed by the stretching. The moves would be designated in the same way as on a rigid chessboard: a7–a5 (conventional chess nomenclature for a pawn moving a particular two squares forward) would still be a7–a5 however the sheet was stretched, but the distances the pawn had to move *would* change. Figure 7 illustrates these comoving flat coordinates.

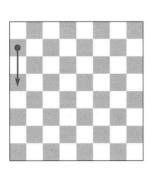 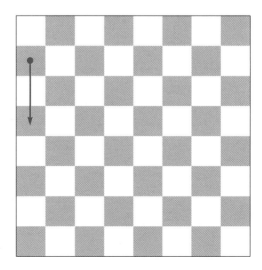

Figure 7   If a chessboard is enlarged, the comoving coordinates of the pieces on the board remain the same.

For a two-dimensional analogy of the meaning of comoving coordinates in *curved* space we return to the surface of a sphere as discussed in Unit 10 & 11. All points on this surface are equivalent, so something akin to the cosmological principle applies. Imagine a coordinate grid drawn on the surface — just like lines of latitude and longitude. Picturing the sphere as a balloon rather than as a rigid globe, we note that as more air is blown into the balloon the coordinates of, say, a fly standing on the surface do not change (Figure 8). The fly remains standing in a particular square of the latitude–longitude grid, and the whole grid expands (uniformly) around it.

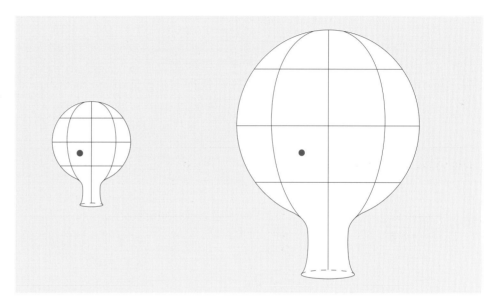

Figure 8   If a grid is drawn on a balloon, the comoving coordinates of points on the balloon remain the same when the balloon is further inflated.

In terms of comoving spherical polar coordinates (θ, ϕ), the metric on such a sphere is given by

$$\Delta l = R(t)\sqrt{(\Delta\theta)^2 + \sin^2\theta\,(\Delta\phi)^2} \tag{1}$$

where $R(t)$ is the radius of the balloon at time t (a time-dependent radius reflecting the effects of the additional air being blown into the balloon). The shortest length between two flies at $(\theta = \pi/2, \phi = 0)$ and $(\theta = \pi/2, \phi = \pi/2)$ is $(\pi/2)R(t)$. This result is obtained by repeating the calculation in SAQ 14(a) of Unit 10 & 11 at a fixed time t. More intuitively, the two flies are joined by a quarter of the equator — i.e. a quarter of a circle of length $2\pi R(t)$. This *increases* as the balloon is inflated even though the (comoving) coordinates of the flies do not change.

Of course, this analogy is not limited to positive curvature — except by the fact that saddle-shaped balloons and horn-shaped balloons (which have negative curvature) are not commonly available!

One further point: Now that we have introduced comoving coordinates, there arises an alternative way of interpreting the Hubble recession. In the previous Unit, we regarded the galaxies as moving through space in the manner in which any other object might move. The redshift exhibited by the galaxies was treated, tentatively, as a Doppler redshift of the sort that might accompany the movement of any source of light.

But now, with comoving coordinates in mind, we can think of *space itself as expanding*. In doing so, space, in a manner of speaking, 'carries' the clusters of galaxies along with it. A galaxy cluster three times further away from us than another has three times as much space between it and ourselves, and will thus recede three times further as that space expands.

In asserting that space itself expands, that does *not* imply that *all* distances expand. If that were to be the case, we would not be aware of any expansion at all. (A distance that has increased by a factor 2, measured on a ruler that has also expanded by a factor 2 will still record the same distance.) The distances that do not increase are those associated with bound systems of particles. This applies whether we are thinking of the Sun, or the Solar System, or a galaxy, or even on the largest scale a cluster of galaxies. It is beyond the size of clusters that we believe the binding forces are sufficiently weak that the expansion of space manifests itself. In the analogy we were using of flies on a rubber balloon, as the latter expands, the flies, of course, do not change size. It is the flies that are the analogy of the clusters of galaxies.

Such a way of viewing the nature of the movement of galaxies alters one's understanding of the origin of the cosmological redshift itself. For a cluster locally 'at rest' (relative to the comoving coordinate system), instead of thinking of the wavelength being Doppler-stretched at emission because of a movement of the source through space (that stretched wavelength then travelling to us without further modification), we think of the wavelength as having its usual value at emission — because there is *no* movement of the source *through* space. But from then on, the distances between the peaks and the troughs of the wavetrain become progressively stretched by the expansion of the space occurring during the time of its journey to us. In other words, the wavelength increases. So, although the wavelength was normal at emission, by the time the light reaches us, it has been redshifted.

4.2   The form of the Robertson–Walker metric

It is clear that our task is to write down, in comoving coordinates $(t, \sigma, \theta, \phi)$, the metric that is appropriate to the large-scale structure of spacetime. In other words, we must find an equation that plays the role of Equation 1 for the four-dimensional spacetime of a homogeneous and isotropic universe. This was the problem that Robertson and Walker solved. They found that

Robertson–Walker metric

$$(\Delta S)^2 = c^2(\Delta t)^2 - R^2(t)\left[\frac{(\Delta\sigma)^2}{1 - k\sigma^2} + \sigma^2(\Delta\theta)^2 + \sigma^2\sin^2\theta(\Delta\phi)^2\right] \qquad (2)$$

where:

ΔS is the Lorentz invariant interval between neighbouring events with comoving coordinates $(t, \sigma, \theta, \phi)$ and $(t + \Delta t, \sigma + \Delta\sigma, \theta + \Delta\theta, \phi + \Delta\phi)$. Note that ΔS was referred to in Block 3 as $c\,\Delta\tau$.

scale factor

$R(t)$ is called the *scale factor* and plays a role analogous to the $R(t)$ in Equation 1. It has the dimensions of length.

spatial curvature parameter

k is called the *spatial curvature parameter*, and is dimensionless. You will see in Section 4.4 that k is related to the curvature of space. There we shall find that there are three cases to be distinguished: $k = 0$, $k > 0$, and $k < 0$. It has become customary to put $k = 0$, $+1$, or -1 respectively. This implies no restriction on generality because there is an arbitrariness of scale in the definition both of $R(t)$ and of σ, so the 'slack' involved in changing the value of k can be taken up by compensating changes in these two other parameters.

Objective 2   **SAQ 3**   Demonstrate the truth of that last statement by showing that it is possible to change R and σ in such a way that an arbitrary change in the numerical value of k does not produce a change in ΔS.

The fact that this metric turns out to have a fairly simple form (for example, it involves no cross-product terms like $\Delta t\,\Delta\sigma$ and the *same* function $R(t)$ multiplies all spatial coordinates $\Delta\sigma$, $\Delta\theta$ and $\Delta\phi$) is a consequence of assuming homogeneity and isotropy and of choosing a suitable coordinate system. Equation 2 represents the most general metric for which the spatial distances vary by a common factor dependent on time, and for which the curvature of space is the same everywhere.

Perhaps a more physical way of seeing the influence of homogeneity on the geometry of spacetime is to take a second look at the concept of synchronized clocks. We have already defined Robertson and Walker's time coordinate by means of identical clocks whose readings are all adjusted to zero at some well-defined stage of the Universe's history. But this *definition* does not, by itself, guarantee that the clocks always remain synchronized. Their behaviour depends on the geometric nature of spacetime and hence on the distribution of matter in our Universe. That is why a homogeneous distribution of matter is so special. In this special case, *there is no reason for one of the identical clocks to run faster or slower than any of the others*, so we *can* assume that they all keep in step (a case of symmetry begets symmetry — see Section 3 of Unit 3).

In order to see how this fact is related to the form of the Robertson–Walker metric, imagine sending a light pulse between two clocks, A and B, in two neighbouring galaxy clusters. The emission of the pulse from clock A constitutes the event $\mathscr{E}_A = (t, \sigma, \theta, \phi)$ and the reception of the pulse at clock B constitutes the event $\mathscr{E}_B = (t + \Delta t, \sigma + \Delta\sigma, \theta + \Delta\theta, \phi + \Delta\phi)$. For convenience, we shall assume that the two clocks are much closer to one another than they are to the origin of our coordinate system, so that $\Delta\sigma \ll \sigma$, $\Delta\theta \ll \theta$, $\Delta\phi \ll \phi$, and $\Delta t \ll t$. Then, according to the Robertson–Walker metric, the Lorentz invariant interval between the two events $\mathscr{E}_A$ and $\mathscr{E}_B$ is given by Equation 2. Since the interval between any

Two events on a null geodesic   two events on the world-line of a light ray is zero (see Section 10 of Unit 10 & 11, remembering that $\Delta S = c\,\Delta\tau$), we conclude that the difference in coordinate time between the events $\mathscr{E}_A$ and $\mathscr{E}_B$ is

$$\Delta t = \frac{1}{c}R(t)\left[\frac{(\Delta\sigma)^2}{1 - k\sigma^2} + \sigma^2(\Delta\theta)^2 + \sigma^2\sin^2\theta(\Delta\phi)^2\right]^{1/2}.$$

The important point to notice about this equation is that the expression in square brackets does not change with time — remember that σ, θ and ϕ are comoving coordinates whose values for any given cluster are fixed. It follows that the time interval, Δt, depends on time only via the scale factor, $R(t)$, *which has the same value across the entire Universe*.

So if *two* such time intervals $(\Delta t)_1$ and $(\Delta t)_2$ are measured to be equal at some given coordinate time, t, they will remain equal at all later times. If, for example, a light signal were sent out to a number of observers with fixed Robertson–Walker coordinates, those who received it at the same value on their clocks would also agree about the arrival times of subsequent signals from the same source. And if two clocks are synchronized in the sense mentioned in Section 4.1, they will remain synchronized forever.

We can therefore conclude that:

> In a homogeneous Universe, described by the Robertson–Walker metric, the same value of the time coordinate can be used for all clusters. In this sense, time is universal.

4.3   Reminder of the meaning of curved space

Before describing the geometric properties of the Robertson–Walker metric, we shall briefly recall from Unit 10 & 11 some geometrical properties of a *two*-dimensional curved space. It is easy to visualize such a space by drawing it embedded in normal three-dimensional space. If the two-dimensional space has a constant positive curvature, it comes out as the surface of a sphere. The sign of its curvature can be found without leaving the surface. This can be done by stepping a distance, r, across the surface of the sphere from some chosen initial point. This distance then becomes the radius of a circle drawn on the sphere about the initial point. On measuring the circumference, C, of this circle, it is found that the ratio C/r is *less* than 2π. This demonstrates that the curvature is *positive* (Figure 9(b)). From our privileged position outside the sphere, the reason for this departure from Euclidean geometry is obvious: the elongation of the radius arises because of the bend in the path it takes.

There are also simple two-dimensional surfaces with negative curvature, such as a saddle and a particular type of horn. As shown in Figure 9(c), if we measure the radius, r, and the circumference, C, of a 'circle' (i.e. a curve of constant radius) on such a surface, we find that C/r is *greater* than 2π and this tells us that the curvature is *negative*. In Section 4.4, we shall use these general properties in the particular case of the Robertson–Walker metric.

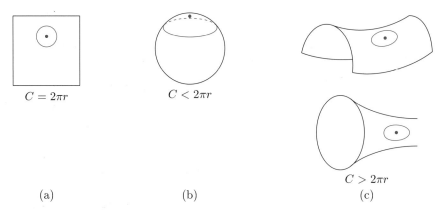

$C = 2\pi r$ $\qquad\qquad$ $C < 2\pi r$

$C > 2\pi r$

(a) $\qquad\qquad$ (b) $\qquad\qquad$ (c)

Figure 9   The ratio of circumference (C) to radius (r) depends on whether the surface has (a) zero curvature, (b) positive curvature, or (c) negative curvature. In each case, the radius r is measured on the surface.

4.4 Geometric properties of the Robertson–Walker metric

The Robertson–Walker metric (Equation 2) is characterized by two parameters: the scale factor, $R(t)$, and the spatial curvature parameter, k. In this Section, we shall investigate the significance of these quantities, and see how the geometry of spacetime is influenced by their values. We shall consider three possibilities in turn:

$$k = 0, \ R(t) = \text{constant} \quad \text{(Section 4.4.1)}$$
$$k = 0, \ R(t) \neq \text{constant} \quad \text{(Section 4.4.2)}$$
$$k = \pm 1 \qquad\qquad\qquad \text{(Section 4.4.3)}$$

4.4.1 Flat spacetime

We set $k = 0$ in Equation 2. Then, in the manner used to solve SAQ 3, we redefine σ so that $R(t)$ (which is here considered to be a constant) becomes $R(t) = 1$. Thus, we obtain the following special form of the Robertson–Walker metric:

$$(\Delta S)^2 = c^2(\Delta t)^2 - [(\Delta\sigma)^2 + \sigma^2(\Delta\theta)^2 + \sigma^2 \sin^2\theta(\Delta\phi)^2].$$

It turns out that the expression in square brackets has a very simple interpretation: it is the square of the distance Δl, in flat Euclidean space, between two points whose spherical polar coordinates are (σ, θ, ϕ) and $(\sigma + \Delta\sigma, \theta + \Delta\theta, \phi + \Delta\phi)$. In other words, if a Cartesian coordinate system were used to relabel the points as (x^1, x^2, x^3) and $(x^1 + \Delta x^1, x^2 + \Delta x^2, x^3 + \Delta x^3)$, the term in square brackets is

$$(\Delta l)^2 = [(\Delta\sigma)^2 + \sigma^2(\Delta\theta)^2 + \sigma^2 \sin^2\theta(\Delta\phi)^2]$$
$$= [(\Delta x^1)^2 + (\Delta x^2)^2 + (\Delta x^3)^2].$$

The proof of this result follows from the definition of $x^1 = \sigma\sin\theta\cos\phi$, $x^2 = \sigma\sin\theta\sin\phi$ and $x^3 = \sigma\cos\theta$. Essentially this derivation is given in Section 6.3 of Unit 10 & 11. You will not be expected to prove it, but you can check that it is reasonable by considering one or two special cases. For example,

if $\quad \Delta\phi = 0, \quad$ then $\quad (\Delta l)^2 = (\Delta\sigma)^2 + \sigma^2(\Delta\theta)^2$

and

if $\quad \Delta\theta = 0 \quad$ and $\quad \theta = \dfrac{\pi}{2}, \quad$ then $\quad (\Delta l)^2 = (\Delta\sigma)^2 + \sigma^2(\Delta\phi)^2$.

These are both expressions for the flat plane in plane-polar coordinates (see Unit 10 & 11, Section 4.4).

From our point of view, it is the *conclusion* that matters: if $k = 0$ and $R(t) = 1$, the Robertson–Walker metric takes the form

$$(\Delta S)^2 = c^2(\Delta t)^2 - [(\Delta x^1)^2 + (\Delta x^2)^2 + (\Delta x^3)^2].$$

Since this is the expression for the interval in the Minkowski metric of *special* relativity, we conclude that, in this case, spacetime is flat.

4.4.2 Flat space, curved spacetime

If $k = 0$ but $R(t) \neq$ constant, i.e. $R(t)$ is now allowed to vary with time, the square of the interval is given by

$$(\Delta S)^2 = c^2(\Delta t)^2 - R^2(t)[(\Delta x^1)^2 + (\Delta x^2)^2 + (\Delta x^3)^2].$$

In this metric, the spatial distance between two simultaneous events is given by the Euclidean formula

$$\Delta l = R(t)\sqrt{(\Delta x^1)^2 + (\Delta x^2)^2 + (\Delta x^3)^2}.$$

Hence, *at any fixed time*, the spatial geometry is Euclidean — the angles of a triangle add up to π radians, and the ratio of the circumference of a circle to its radius is 2π. So space is flat. But *spacetime* is not flat because the scale factor $R(t)$ causes the distance between comoving locations to change from one moment to the next. The analogy you should have in mind is that of the expanding chessboard: at any given moment, its geometry is that of a rigid board, but as time passes you can easily see that something strange is happening. In the case of the Robertson–Walker metric, it is the dynamics of particles that reveals the curvature of spacetime. For example, if particles are projected directly at each other, they will never collide if $R(t)$ increases rapidly enough!

4.4.3 Curved space

Let us now take the cases where $k = +1$, or -1. In order to simplify matters, we shall consider the curvature of *space* rather than of spacetime. That is, we shall consider Equation 2 at an instant of coordinate time, t, so that $\Delta t = 0$, and write

$$(\Delta S)^2 = -R^2(t)\left[\frac{(\Delta\sigma)^2}{1 - k\sigma^2} + \sigma^2(\Delta\theta)^2 + \sigma^2\sin^2\theta(\Delta\phi)^2\right] \equiv -(\Delta l)^2$$

where Δl is an interval of what was called 'proper length' in Unit 10 & 11.

Now imagine a giant circle (Figure 10), whose points have coordinates

$$(\sigma_0, \theta = \pi/2, \phi)$$

where $0 < \phi \leq 2\pi$. We can calculate the circumference of this circle by the methods of Unit 10 & 11.

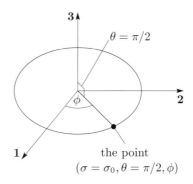

$\theta = \pi/2$

the point
$(\sigma = \sigma_0, \theta = \pi/2, \phi)$

Figure 10

We begin by introducing a parametric representation for the circle:

$$\left.\begin{array}{l} \sigma(s) = \sigma_0 \\ \theta(s) = \pi/2 \\ \phi(s) = s \end{array}\right\} \quad 0 < s \leqslant 2\pi.$$

Recall that the dot here signifies a derivative with respect to parameter s.

Then, adapting the formula for the length of a curve on a sphere (Equation 29 of Unit 10 & 11), to the case of a curve in three dimensions, we find for the circumference, C, of the circle

$$C = \int_{s=0}^{s=2\pi} R(t)\left[\frac{\dot{\sigma}^2}{1 - k\sigma^2} + \sigma^2\dot{\theta}^2 + \sigma^2\sin^2\theta\dot{\phi}^2\right]^{1/2}\,\mathrm{d}s$$

$$= \int_{s=0}^{s=2\pi} R(t)\left[0 + 0 + \sigma_0^2\sin^2\frac{\pi}{2} \times 1^2\right]^{1/2}\,\mathrm{d}s$$

$$= R(t)\sigma_0 2\pi.$$

A similar calculation can be performed for the radius, r, of the circle. This time we use the parametric representation for the radius:

$$\left.\begin{array}{l} \sigma(s) = s \\ \theta(s) = \pi/2 \\ \phi(s) = \phi_0 \end{array}\right\} \quad 0 < s \leqslant \sigma_0.$$

It follows that

$$r = \int_{s=0}^{s=\sigma_0} R(t) \left[\frac{\dot{\sigma}^2}{1 - k\sigma^2} + \sigma^2\dot{\theta}^2 + \sigma^2 \sin^2\theta\dot{\phi}^2 \right]^{1/2} \mathrm{d}s$$

$$= \int_{s=0}^{s=\sigma_0} R(t) \left[\frac{1}{1 - ks^2} + 0 + 0 \right]^{1/2} \mathrm{d}s$$

$$= R(t) \int_{s=0}^{s=\sigma_0} \frac{\mathrm{d}s}{\sqrt{1 - ks^2}}.$$

In the special case $k = 0$, we recover

$$r = R(t) \int_{s=0}^{s=\sigma_0} 1 \times \mathrm{d}s = R(t)\sigma_0$$

so that

$$\frac{C}{r} = \frac{R(t)\sigma_0 2\pi}{R(t)\sigma_0} = 2\pi$$

as one would expect for a flat space.

However, if $k = +1$, the integrand $\dfrac{1}{\sqrt{1 - ks^2}}$ is $\dfrac{1}{\sqrt{1 - s^2}}$ and

$$r = R(t) \sin^{-1}\sigma_0$$

and this means r is *greater* than $R(t)\sigma_0$. See Figure 11.

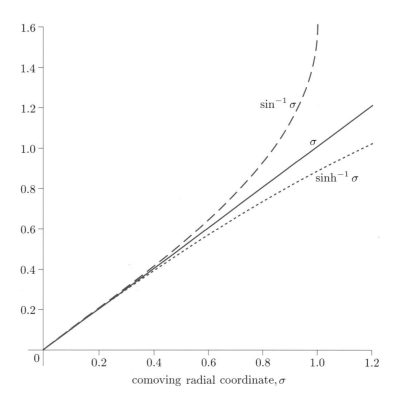

Figure 11 A reminder of the behaviour of the $\sin^{-1}$ and $\sinh^{-1}$ functions.

Hence

$$\frac{C}{r} < 2\pi \quad \text{for } k = +1. \tag{3}$$

Similarly, if $k = -1$, the integrand becomes $\dfrac{1}{\sqrt{1 + s^2}}$ and we have

$$r = R(t) \sinh^{-1} \sigma_0$$

and this means r is *less* than $R(t)\sigma_0$. Hence

$$\frac{C}{r} > 2\pi \quad \text{for } k = -1. \tag{4}$$

We conclude that the curvature of space is dependent on the parameter k: if $k = +1$, space has a positive curvature (like a sphere); if $k = -1$, space has a negative curvature (like a trombone horn). On the other hand, the radius and circumference are also proportional to $R(t)$, which again illustrates the role of $R(t)$ in giving a length scale at a given time.

Although the value of k (i.e. 0, +1, or –1) is crucial in determining the *type* of curvature of space, you should realize that we are *not* claiming that k by itself gives the *value* of the spatial curvature. In fact it is the expression $k/R^2(t)$ that gives the measure of the non-Euclidean nature of space.

An interesting consequence of this is that, if $R(t)$ were to increase with time, the value of $k/R^2(t)$ would progressively decrease, and the departure from Euclidean geometry would become less pronounced. This can be understood by remembering the two-dimensional analogy of the inflated balloon. The 'curvature' of the surface of the balloon is defined as $1/a^2$, where a is the radius of the balloon. This clearly becomes less as the balloon expands.

Thus, there are two ways of having a flat space: Either it is *exactly* flat because $k = 0$, or it *approaches* flatness through R^2 being very large compared to k (which has the value +1 or −1). Note that by 'very large' we mean '*exceedingly* large'. For example, normally one might regard an R^2 of 10^6m^2 as very large, but this would correspond to $R = 10^3$ m, or a radius of only 1 kilometre for the Universe. Clearly we have something much larger than *that* in mind!

The question of the actual value of k for the space of our Universe (whether it is 0, +1, or −1) is deferred until the next Unit. Meanwhile, the remainder of this Unit will concentrate on the scale factor $R(t)$.

Objectives 2 and 3   **SAQ 4**   Which of the following statements are true for an isotropic, homogeneous universe described by the Robertson–Walker metric?

(a) The spatial coordinates, $(\sigma_1, \theta_1, \phi_1)$ and $(\sigma_2, \theta_2, \phi_2)$, of two given galaxy clusters do not change with time.

(b) The time taken for a light signal to pass between two given galaxy clusters is the same at all phases of the universe.

(c) If $k = 0$, spacetime is necessarily flat.

(d) If $k = -1$, the circumference of a circle is shorter than 2π times its radius.

(e) If $R(t) \neq 1$ at all times, space cannot be Euclidean.

Objective 2

SAQ 5 Observers A, B and C live on planets in different clusters of galaxies.

A sends out a radio signal, which is reflected by B and C; the replies are received simultaneously by A.

B sends out a radio signal, which is reflected by A and C; the replies are received simultaneously by B.

C sends out a radio signal, which is reflected by A and B; the replies are received simultaneously by C.

Observer A then measures the angles α between the lines of sight to B and C.

(a) What can be deduced about the spatial curvature parameter k if $\alpha = 60°$?

(b) What can be deduced if $\alpha < 60°$?

(c) What can be deduced if $\alpha > 60°$?

Objective 4

SAQ 6 'The metric used in special relativity is the particular case of the Robertson–Walker metric for which $k = 0$, i.e. for which space is flat.' Comment on the accuracy of this statement.

5 The Hubble parameter

In this section, we show that in any universe in which $R(t)$ increases with time, distant galaxies will exhibit redshifts with the same characteristics as those observed in our Universe. It is fairly straightforward to find the relationship between the observed value of H and the scale factor $R(t)$ by the following argument:

As a first step, we shall use the Robertson–Walker metric and try to compute the cosmological redshift, z, of light from a distant galaxy in terms of the scale factor $R(t)$. We are free to choose any comoving origin because of the homogeneity of the Universe, and for simplicity we shall choose the Earth. We consider light rays travelling radially inwards. For radial motion $\Delta\theta = \Delta\phi = 0$. For light, $\Delta S = 0$. Thus

$$0 = (\Delta S)^2 = c^2(\Delta t)^2 - R^2(t)\frac{(\Delta\sigma)^2}{1 - k\sigma^2}. \tag{5}$$

> The proper motions of the Earth round the Sun, and of the Solar System around our Galaxy, and of our Galaxy in our Local Group must be subtracted out, as mentioned previously.

Light leaving a galaxy at (σ_1, θ, ϕ) at time t_1 reaches the Earth at the (later) time t_0. Using the square root of Equation 5 we can write

$$\int_{t_1}^{t_0} \frac{c\,dt}{R(t)} = \int_0^{\sigma_1} \frac{d\sigma}{\sqrt{1 - k\sigma^2}}. \tag{6}$$

Reminder regarding notation:

By convention, the suffix $_0$ when applied to the time, t_0, (or to the value of the Hubble parameter, H_0) always refers to the value at the *present* time. This can lead to a situation (as we have here) where t_1 refers to an *earlier* time. As noted in the previous Unit, such a reversal of order is counter-intuitive, but has become the accepted custom.

The interpretation of this equation is made much easier by the fact that it is written in comoving coordinates. The value of the comoving coordinate, σ_1, for the distant galaxy is constant. So we can immediately write down the corresponding equation for light which leaves the galaxy at a later time, $t_1 + \Delta t_1$. It will arrive at our telescope a little after the time t_0, let us say at time $t_0 + \Delta t_0$. Here, Δt_0 is *not* the same as Δt_1 (it is, in fact, their ratio which we are trying to find). We write:

$$\int_{t_1+\Delta t_1}^{t_0+\Delta t_0} \frac{c\,dt}{R(t)} = \int_0^{\sigma_1} \frac{d\sigma}{\sqrt{1 - k\sigma^2}}. \tag{7}$$

It is the fact that the right-hand side remains the same that enables us to calculate Δt_0. Subtracting Equation 6 from Equation 7 gives

$$\int_{t_1+\Delta t_1}^{t_0+\Delta t_0} \frac{c\,dt}{R(t)} - \int_{t_1}^{t_0} \frac{c\,dt}{R(t)} = 0.$$

Now there is a range which is common to these two integrals, as shown by the shaded area in Figure 12. The common part cancels out, leaving

$$\int_{t_0}^{t_0+\Delta t_0} \frac{c\,dt}{R(t)} - \int_{t_1}^{t_1+\Delta t_1} \frac{c\,dt}{R(t)} = 0.$$

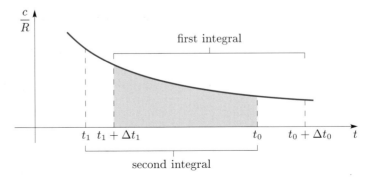

Figure 12   The common part of the integrals, for an expanding universe.

This equation is true for any length of time Δt_1, but we have written it in the form of a small time increment because what we are really interested in is how the frequency of radiation is affected. To be specific, let Δt_1 be the period of oscillation of light in the first line of the hydrogen spectrum, which has a frequency of 2.47×10^{15} Hz, so that $\Delta t_1 = 4.05 \times 10^{-16}$ s. Over this time, the change in $R(t)$ is completely negligible, so we can drop the integral signs and write

$$\frac{c(t_0 + \Delta t_0 - t_0)}{R(t_0)} = \frac{c(t_1 + \Delta t_1 - t_1)}{R(t_1)}$$

so that

$$\frac{c\,\Delta t_0}{c\,\Delta t_1} = \frac{R(t_0)}{R(t_1)}. \tag{8}$$

We have left in the factor c because it makes it easier to return to the parameter which is actually measured by the astronomer — the wavelength of light, which is $c \times$ period. Thus

$$\frac{\lambda_0}{\lambda_1} = \frac{R(t_0)}{R(t_1)} \tag{9}$$

where λ_0 and λ_1 are the observed and emitted wavelengths, respectively. Since $\Delta\lambda \equiv \lambda_0 - \lambda_1$ and the redshift z is defined as $\Delta\lambda/\lambda_1$,

$$z = \frac{\lambda_0 - \lambda_1}{\lambda_1} = \frac{\lambda_0}{\lambda_1} - 1 \tag{10}$$

$$z = \frac{R(t_0)}{R(t_1)} - 1. \tag{11}$$

This formula gives us a way of interpreting the cosmological redshift, z, in terms of the scale factor $R(t)$ in the Robertson–Walker metric.

From the form of this equation, it is clear that z must be finite for all galaxies, since they were certainly formed when $R(t_1)$ was non-zero.

In our Universe, the visible light from distant galaxies is shifted towards the red end of the spectrum — towards longer wavelengths. It follows that, in Equation 11, $z > 0$ and $R(t_0) > R(t_1)$. We conclude that the scale factor $R(t)$ is increasing as the Universe gets older. In loose terms, one might say that since the whole of space is expanding, the wavelength of the light emitted from a galaxy is steadily stretched out during its long journey to us on Earth. This you recall was one of the interpretations of the redshift offered at the end of Section 4.1.

Objective 5   **SAQ 7**   A certain quasar has a redshift $z = 4.5$. By what factor has the Universe expanded since the quasar emitted the light we receive today?

Objective 5   **SAQ 8**   Show that $z < 0$ for a contracting universe.

Objective 5   **SAQ 9**   Draw the equivalent of Figure 12 for a contracting universe.

If the Robertson–Walker metric really does describe the large-scale structure of spacetime, we can use Equation 11 to relate the observationally measured values of z to the values of $R(t)$ appearing in the metric. However, it is important to realize that the measurements of z

do not give the *absolute value* of $R(t)$, but only the *ratio* of two values at different times. Indeed, for most purposes the key quantities are $\dfrac{1}{R}\dfrac{dR}{dt}$ and $\dfrac{1}{R}\dfrac{d^2R}{dt^2}$, which are both invariant if R is multiplied by an arbitrary constant. Or expressing that in words: what matters is the *fractional* rate of change of R, and the rate at which that fractional change itself varies with time.

In order to clarify the connection between theory and observation, we need to re-express Equation 11 in terms of these two rates of change. We can do this as follows. We use a standard mathematical approximation, known as the Taylor expansion, to write the value of R at some earlier time, t_1, in terms of the value of R at the present time, t_0:

$$R(t_1) = R(t_0) + \left(\frac{dR}{dt}\right)_{t=t_0}(t_1 - t_0) + \frac{1}{2}\left(\frac{d^2R}{dt^2}\right)_{t=t_0}(t_1 - t_0)^2$$
$$+ \text{ higher-order terms.}$$

It does not matter if you are unfamiliar with this Taylor expansion. The only point of importance is that $R(t)$ can be approximated by a sum of powers of $(t_1 - t_0)$, including constant, linear, quadratic and higher-order terms.

Factoring out $R(t_0)$ and neglecting the higher-order terms, we get

$$R(t_1) = R(t_0)\left[1 + \left(\frac{1}{R}\frac{dR}{dt}\right)_{t=t_0}(t_1 - t_0) + \frac{1}{2}\left(\frac{1}{R}\frac{d^2R}{dt^2}\right)_{t=t_0}(t_1 - t_0)^2\right]. \quad (12)$$

We can express the factor $\left(\dfrac{1}{R}\dfrac{dR}{dt}\right)_{t=t_0}$ in a more familiar fashion in the following way: Equation 11 can be written

$$z = \frac{R(t_0) - R(t_1)}{R(t_1)}.$$

$\rho 22$

But from Equations 4 and 6 of Unit 13, we have (for small r)

$$z = Hr/c$$

where r is the distance to the galaxy and is defined for small $(t_0 - t_1)$ as

$$r = c(t_0 - t_1).$$

Substituting for z and r we obtain

$$H = \frac{R(t_0) - R(t_1)}{R(t_1)(t_0 - t_1)}.$$

i.e. $H = \dfrac{1}{R}\dfrac{\Delta R}{\Delta t}.$

Hubble parameter
(new definition)

In the limit as Δt approaches zero, we have $H = \dfrac{1}{R}\dfrac{dR}{dt}$, so

$$H_0 = \left(\frac{1}{R}\frac{dR}{dt}\right)_{t=t_0} \quad (13)$$

where we have included a zero subscript to indicate that this is the Hubble parameter evaluated at the present time. Thus we find one of the terms in Equation 12 relates to the familiar observable, the *Hubble parameter*.

deceleration parameter

As regards the next term in Equation 12, cosmologists go on to define a further quantity (a dimensionless one) called the *deceleration parameter*, $q(t)$:

$$q(t) \equiv -\frac{1}{H^2}\left(\frac{1}{R}\frac{d^2R}{dt^2}\right) = -\frac{R\dfrac{d^2R}{dt^2}}{\left(\dfrac{dR}{dt}\right)^2} \quad (14)$$

which in the particular case $t = t_0$ becomes

$$q_0 \equiv q(t_0) = -\frac{1}{H_0^2}\left(\frac{1}{R}\frac{\mathrm{d}^2 R}{\mathrm{d}t^2}\right)_{t=t_0} = -\left(\frac{R\frac{\mathrm{d}^2 R}{\mathrm{d}t^2}}{\left(\frac{\mathrm{d}R}{\mathrm{d}t}\right)^2}\right)_{t=t_0}. \tag{14a}$$

Equation 12 then takes the form

$$\frac{R(t_0)}{R(t_1)} = (1 + H_0(t_1 - t_0) - \tfrac{1}{2}q_0 H_0^2(t_1 - t_0)^2)^{-1}.$$

The left-hand side of this equation is simply $z + 1$, and the right-hand side can be expanded using the series expansion $(1 + x)^{-1} = 1 - x + x^2 - x^3 \ldots$. Doing this and neglecting all terms $(t_1 - t_0)^n$ for which n is greater than 2, we get:

$$\frac{R(t_0)}{R(t_1)} - 1 = z = H_0(t_0 - t_1) + H_0^2\left(1 + \frac{q_0}{2}\right)(t_0 - t_1)^2.$$

Recalling that $r = c(t_0 - t_1)$, we get

$$z = H_0\frac{r}{c} + \left(1 + \tfrac{1}{2}q_0\right)\left(H_0\frac{r}{c}\right)^2. \tag{15}$$

Under certain circumstances, it is possible to approximate still further. The best estimate of the current value of the Hubble parameter is so tiny ($H_0 \approx 2 \times 10^{-18}\,\mathrm{s}^{-1} \approx 0.5 \times 10^{-10}\,\mathrm{yr}^{-1}$), that $H_0 r/c \ll 1$ for any galaxy whose distance from the Earth is much less than 10^{10} light-years. So, for all but the most distant observable galaxies, we can neglect the term in $(H_0 r/c)^2$. Therefore

$$z \approx H_0\frac{r}{c} \tag{16}$$

which is the familiar 'small r' relation we used a little earlier, and which leads to Hubble's law ($v = H_0 r$), when the redshift is interpreted as a Doppler shift. Now, however, we have, through Equation 15, derived a correction to the relation which might give rise to observable effects for more distant galaxies, perhaps yielding a value of the deceleration parameter q_0. This is very important for cosmology, as we shall see.

Note that Equation 15, though a better approximation than Equation 16, still cannot be used where z is large, as it is for the distant quasars, and for the cosmic microwave background for which $z \approx 1000$.

However, you should note that the statement of Hubble's law itself (the expression involving speed rather than redshift) *is* exact. We can see this by considering the proper distance, D_p, to a distant galaxy, and the rate at which it changes. According to the Robertson–Walker metric

Recall that 'proper distance' was discussed in Unit 10 & 11, Section 9.1.

$$D_\mathrm{p} = R(t)\int_0^\sigma \frac{\mathrm{d}\sigma}{\sqrt{1 - k\sigma^2}}$$

so that

$$\frac{\mathrm{d}D_\mathrm{p}}{\mathrm{d}t} = \frac{\mathrm{d}R}{\mathrm{d}t}\int_0^\sigma \frac{\mathrm{d}\sigma}{\sqrt{1 - k\sigma^2}} = \frac{1}{R}\frac{\mathrm{d}R}{\mathrm{d}t}D_\mathrm{p}.$$

Thus

$$v = H_0 D_\mathrm{p}$$

which is Hubble's law as applied to the galaxy under consideration, and, in the case of the Robertson–Walker metric, is, as we see, an exact expression.

It is worth mentioning at this juncture that the videocassette accompanying this Unit has a sequence devoted to the motion of 'cotton wool' clusters of galaxies stuck onto an expanding balloon. The equation above is an exact mathematical representation of that situation. When you look at the balloon from the outside, you can immediately see that the rate of increase of the 'proper distance' between two galaxies (measured in the surface of the balloon) is proportional to the 'proper distance' between them — a property of the Robertson–Walker metric as demonstrated here.

To summarize, Equation 16 is a first-order approximation for the redshift, z. A second-order term, which depends on q_0, is expected to be present. The effect of the second-order term in Equation 15 obviously increases with the value of r. The next question, therefore, is whether observations at large distances reveal the effect of, and hence the value of, q_0.

In practice it is very difficult to determine q_0 from direct observations of galaxies. Figure 13 shows some well known recent results (1999) based on Type Ia supernovae. (You will recall from Unit 13 that supernovae of this kind provide a very bright 'standard lamp'.) The dashed lines show the expected relationship between the plotted quantities in a universe described by the Robertson–Walker metric for various values of the deceleration parameter q_0. These data seem to favour a negative value of q_0, between -0.25 and -1, though they are hard to read. Similar data, gathered more recently, or by independent observers, also indicate a negative value for q_0. However, the real issue is whether to place much confidence in any of these results.

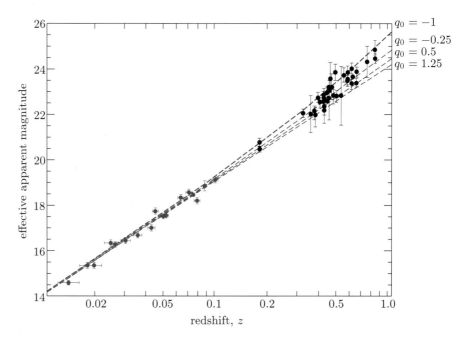

Figure 13   A plot of effective apparent magnitude against redshift for Type Ia supernovae. The dashed lines show the expected correlation between these two quantities in a universe described by a Robertson–Walker metric for the indicated values of q_0. The figure has been adapted from one produced by the Supernova Cosmology Project.

These particular data, along with the other similar results, have been hotly debated by astronomers and observational cosmologists. The quality of the observational work is not in question. The problem, as the observers themselves recognize, concerns the interpretation and arises mainly from the extreme distance of the faintest of these supernovae. The most distant, with redshifts of about $z = 1$ are so far away that the light from them has taken a very long time to reach the Earth. The concern is that these very ancient Type Ia supernovae may have been somewhat different from their present-day counterparts. If so, the departure from linearity indicated by the data points in Figure 13 may be the result of stellar or galactic evolution rather than a true measure of the deceleration parameter.

Support for a negative value of q_0 is, however, available from another source. The attempts to model the angular power spectrum of the cosmic microwave background, described in Unit 13 Section 6.3, also indicate $q_0 < 0$. The value indicated by the Wilkinson Microwave Anisotropy Probe results released in 2003 is $q_0 \simeq -0.6$.

Objective 6 **SAQ 10** (a) Show that

$$q = -1 - \frac{1}{H^2}\frac{dH}{dt}$$

is an identity which follows from the definitions of q and H.

(b) One might expect that the effect of gravity would be to slow down (i.e. decelerate) the rate of expansion of the Universe. Assuming that this is the case, so that $q > 0$, show that the Hubble parameter cannot increase.

Objective 6 **SAQ 11** Briefly list the steps in the argument given in this section that lead eventually to a relationship between redshift and distance that includes a second-order term.

Although the Universe is currently expanding, implying that $R(t)$ is increasing and $H_0 > 0$, SAQ 10 shows that if the current value of the deceleration parameter is positive, i.e. $q_0 > 0$, then the Hubble parameter cannot increase. An even more direct result concerns the case $q_0 < 0$. If q_0 really is negative, as recent results appear to indicate, then it follows from the definition of q_0 that $d^2 R/dt^2$ must currently be positive. This, of course, means that the rate at which $R(t)$ is increasing is itself increasing. In other words the Universe is not only expanding, but it is expanding at a rate that is currently accelerating.

If, as SAQ 10 suggests, the gravitational effect of matter is to slow down cosmic expansion, you might well ask what is causing the acceleration. Even if future developments overturn the currently popular notion of an accelerating universe, you might still ask whether there is any mechanism within relativistic cosmology that can account for an accelerating expansion. These are questions that will both be addressed in Unit 15.

6   Direct measurement of R?

We have been able to interpret the cosmological redshift in terms of the scale factor R according to Equation 11:

$$z = \frac{R(t_0)}{R(t_1)} - 1.$$

Can this relationship be tested by a direct measurement of R? It would indeed be very useful if an experiment could be devised which measured R by methods which were independent of the astronomical data that are used to define H_0. Unfortunately no such experiment, has yet been devised.

Nevertheless, the question of what it would entail to measure the scale factor R in the laboratory is an illuminating one. The fact that the scale change $R(t_0)/R(t_1)$ shows up as an astronomical redshift is due to the very special circumstance that the light whose frequency is to be measured has been 'stored' a long time while it travelled through space to us. Could this circumstance not be replicated in the laboratory?

Let us suppose that we have a pair of mirrors kept at a fixed separation by a rigid connecting rod. The mirrors face each other. They have been polished with such surpassing skill that light can continue bouncing backwards and forwards between them for a year. In response to the change in R, the wavelength of the light ought, over a period of time, to increase relative to the fixed separation of the mirrors. Could we *in principle* find any frequency change during this time?

No. Because the aim of the experiment is to find R — a parameter that describes the geometry of spacetime on the galactic scale — it is essential for the mirror apparatus to be in free fall, so that its geodesic, and the geodesic of the light ray bouncing between the mirrors, may reveal the nature of the spacetime. If the mirrors were, say, bolted down in the laboratory, we would learn nothing about R because non-gravitational forces would come into play and the effects of Einstein's theory of general relativity would be masked.

Even having one's apparatus in free fall is not by itself sufficient. One must also choose with care the region of spacetime to test. If the apparatus were to be in free fall near the Earth, or in the Solar System, or even within our Galaxy, it would not reveal anything about R. The Robertson–Walker metric describes the nature of spacetime that applies on the largest scales. In the vicinity of large aggregates of matter there are relatively large local distortions of spacetime. A measurement made within our Galaxy could not reveal the scale factor R — any more than a survey of the irregularities on your living room floor could reveal the radius of the Earth. For this reason it is doubtful whether we, or our descendants, will ever be able to determine R directly.

It remains to ask whether there is any *theoretical* argument, based on general relativity, that can be used to *predict* the behaviour of R. It turns out that there is. The scale factor R describes the curvature of spacetime. But we know from Unit 10 & 11 that curvature is related, via Einstein's field equations, to the distribution of mass. In the next Unit, we shall discuss the predictions concerning R that can be made from Einstein's field equations. We shall then be in a position to consider some fascinating questions concerning the history and fate of our Universe.

Summary

1 There are inhomogeneities in the distribution of galaxies up to the scale of superclusters. However, for the purposes of understanding the underlying structure of the Universe as a whole, it is regarded as reasonable to assume that, above that scale, the distribution is isotropic and homogeneous. This hypothesis is known as the cosmological principle.

2 The clusters are sufficiently far apart for them to be regarded as freely-falling point particles, the geodesics of which reveal the underlying spacetime structure of the Universe.

3 For an isotropic and homogeneous Universe, spacetime can be described by the Robertson–Walker metric (Equation 2).

4 The Robertson–Walker metric makes use of the comoving coordinate σ. This is a dimensionless radial coordinate, the value of which remains constant throughout time for any particular cluster. The origin of the coordinates can be any point in space.

5 The Robertson–Walker metric assumes that the same value of the time coordinate t can be used for all clusters.

6 The curvature of spacetime is characterized by two parameters: a scale factor, R; and a spatial curvature parameter, k. The scale factor R has the dimensions of length, the curvature parameter is dimensionless and may be taken to be -1, 0 or $+1$.

7 For $k = 0$, we distinguish between flat spacetime, and a flat space coupled to a curved spacetime. For $k = +1$, the curvature of space is positive (like a sphere); alternatively, if $k = -1$, the curvature of space is negative (like a gramophone horn).

8 In each case, the curvature of space is k/R^2. From this it can be seen that the curvature of space can be *approximately* flat for the $k = +1$ and $k = -1$ cases, provided that R^2 is sufficiently large.

9 The redshift in the spectral lines of light from galaxies finds a natural explanation in the expansion associated with an increasing value for R, according to Equation 11. Using the relativistic picture, it is clear that the right-hand side of the relationship, $z = H_0 r/c$, is just the first term in a series involving higher-order terms, H_0 being the value of the Hubble parameter evaluated at $t = t_0$, and expressed in terms of R by Equation 13.

10 The second-order term involves the deceleration parameter, q_0. This is a measure of the slowing down of the expansion of the Universe.

11 From the currently available data, it is difficult to estimate the value of q_0. However, recent measurements indicate that q_0 is negative, implying that the expansion of the Universe is accelerating.

12 The Robertson–Walker metric implies that the rate of increase in proper distance between galaxy clusters is proportional to the proper distance. This can be modelled by the behaviour of objects stuck to an expanding balloon.

13 It is impossible to directly measure the scale factor R of the Universe from local experiments in the Solar System — or, for that matter, from anywhere within the Galaxy.

Band 4 of AC4 comments on this Unit.

Acknowledgements

Grateful acknowledgement is made to the following sources for material used in this Unit:

Figure 1: from Le Grand Atlas de L'Astronomie (1983) by permission of Encyclopaedia Universalis; *Figure 2*: courtesy of R. Brent Tully/Sky & Telescope; *Figure 3*: courtesy of S. J. Maddox, G. Efstathiou and W. J. Sutherland, University of Oxford; *Figure 4*: courtesy of Robert Williams and the Hubble Deep Field Team (STScI) and NASA; *Figure 5*: courtesy of Professor Richard Ellis, University of Cambridge, Institute of Astronomy; *Figure 6*: based on G. G. Pooley and M. Ryle (1968) in *Monthly Notes of the Royal Astron. Soc.*, **139**, 515; *Figure 13*: S. Perlmutter *et al* (1999), Measurements of Omega and Lambda from 42 -Redshift Supernovae, *The Astrophysical Journal*, **517**, 1999, University of Chicago Press.

Self-assessment questions — answers and comments

foreground galaxies
belonging to one's own cluster

other clusters
in background

Figure 14 (For SAQ 1) The sky as seen from a point in a cluster of galaxies.

SAQ 1 The view will be roughly symmetrical, so we need only sketch the view in one direction as in Figure 14. There will be other galaxies of the same cluster in the foreground at various distances, and a fairly clear view to other clusters in the background; the distant clusters will be fairly evenly distributed.

SAQ 2 Assume each galaxy has a luminosity L, so that a galaxy at distance r has a flux density, S, proportional to L/r^2. We write $S = wL/r^2$, where w is a constant.

Then the galaxies with brightness greater than or equal to S are those within a radius a, where $a = \sqrt{wL/S}$.

The volume of a sphere with radius a is

$$V = \frac{4}{3}\pi \left(\sqrt{\frac{wL}{S}} \right)^3 = \frac{4}{3}\pi \left(\frac{wL}{S} \right)^{3/2}.$$

In a homogeneous universe, the number of galaxies per unit volume, n, is constant. We therefore conclude that the number with brightness greater than S is

$$nV = n\frac{4\pi}{3} \left(\frac{wL}{S} \right)^{3/2} \propto S^{-3/2}$$

which establishes the $S^{-3/2}$ power law.

SAQ 3 The aim is to change the value of k without altering the Lorentz invariant interval ΔS. To be specific, suppose we decided to increase the value of k by a factor of 2, then the denominator of the fraction in which k appears in Equation 2 can be kept unchanged by decreasing the value of σ^2 by the factor 0.5. This will mean the values of $\Delta \sigma^2$ and σ^2 appearing in the three terms within the square brackets will be decreased by a factor 0.5. This can be compensated for by increasing R^2 by a factor 2. Thus we have changed the value of k without altering the spatial contribution to the value of ΔS.

SAQ 4 Statement (a) is true and all the others are false. (If you have difficulty in understanding this answer you should reread Section 4, noting the distinction between flatness of *space* and flatness of *spacetime*.)

SAQ 5 According to the measurements of the three observers, the triangle ABC has

$$AB = AC, \quad BA = BC, \quad CA = CB.$$

We can therefore conclude that the triangle is equilateral, and that the angles subtended at A, B and C are equal. Hence the sum of the angles in ABC is 3α.

(a) If $\alpha = 60°$, the sum of the angles in the triangle is 180°, so space is flat and $k = 0$.

(b) If $\alpha < 60°$, the sum of the angles in the triangle is *less* than 180°, so space has negative curvature and $k = -1$.

(c) If $\alpha > 60°$, the sum of the angles in the triangle is *greater* than 180°, so space has positive curvature and $k = +1$.

SAQ 6 The metric used for special relativity differs in that it does not feature the $R(t)$ factor. It is true that $k = 0$ for both cases, and this means space is flat. But the presence of the scale factor, $R(t)$ in the Robertson–Walker metric allows *spacetime* to be non-flat.

SAQ 7 Substituting $z = 4.5$ in Equation 10, it is clear that $R(t_0)/R(t_1) = 5.5$.

SAQ 8 If the Universe is contracting, the present value of R is less than the value of R at an earlier time, so that $R(t_0)/R(t_1) < 1$. Thus $z < 0$.

SAQ 9 The equivalent of Figure 12 for a contracting universe is shown in Figure 15.

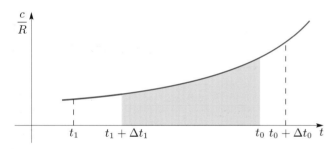

Figure 15 (For SAQ 9) The common part of the integrals for a contracting universe.

SAQ 10 (a) $H \equiv \dfrac{\mathrm{d}R}{\mathrm{d}t} \times \dfrac{1}{R}$.

Therefore $\dfrac{\mathrm{d}H}{\mathrm{d}t} = \dfrac{\mathrm{d}^2 R}{\mathrm{d}t^2} \times \dfrac{1}{R} - \dfrac{1}{R^2}\left(\dfrac{\mathrm{d}R}{\mathrm{d}t}\right)^2$.

Rearranging, $\dfrac{\mathrm{d}^2 R}{\mathrm{d}t^2} = R\left[\dfrac{\mathrm{d}H}{\mathrm{d}t} + \dfrac{1}{R^2}\left(\dfrac{\mathrm{d}R}{\mathrm{d}t}\right)^2\right]$.

But $q \equiv -\dfrac{\dfrac{\mathrm{d}^2 R}{\mathrm{d}t^2} \times R}{\left(\dfrac{\mathrm{d}R}{\mathrm{d}t}\right)^2}$.

Substituting for $\dfrac{\mathrm{d}^2 R}{\mathrm{d}t^2}$,

$$q = -\dfrac{R^2 \dfrac{\mathrm{d}H}{\mathrm{d}t} + \left(\dfrac{\mathrm{d}R}{\mathrm{d}t}\right)^2}{\left(\dfrac{\mathrm{d}R}{\mathrm{d}t}\right)^2} = -1 - \dfrac{1}{H^2}\dfrac{\mathrm{d}H}{\mathrm{d}t}.$$

(b) Because q is positive and so is H, $\mathrm{d}H/\mathrm{d}t$ must be negative (i.e. the Hubble parameter cannot increase).

SAQ 11 The essential steps are:

(a) assume that the distant cluster has constant comoving coordinates;

(b) choose the geodesic of a light ray;

(c) integrate the metric to find the time to the Earth for two different emission times;

(d) take the difference to find the dilation of the time-scale;

(e) convert to the dilation of the wavelength;

(f) express the result in terms of the astronomer's redshift parameter, z.

Unit 15 The evolution of the Universe

*Prepared by the Course Team
with B.F. Schutz*

Contents

Aims

In this Unit we intend to:

1 Introduce the Friedmann equations of the Friedmann–Robertson–Walker cosmology, and show how they are used to understand the structure and evolution of the Universe;

2 Describe our current understanding of the history of the Universe, from shortly after its beginning to the present day, and show how its future might be predicted.

Objectives

When you have finished studying this Unit, you should be able to:

1 Explain the significance of the terms in the Friedmann equations for an isotropic, homogeneous universe (without recalling the equations), and distinguish between observational quantities and the theoretical parameters of the model.

2 Deduce and describe the differences between open and closed universes.

3 Derive an upper limit for the age of the Universe from the current value of the Hubble parameter, explaining why it is an upper limit, and compare it with other estimates relating to the age of the Universe.

4 Derive the time dependence, $R(t) = At^{2/3}$, of a spatially flat, matter-dominated universe, and apply it to observations of our Universe.

5 Explain how the mass density of different regions of the Universe is estimated, and why astronomers believe that most of the mass of the Universe is unseen.

6 Give an account of the process of inflation that might have occurred in the early development of the Universe.

7 Describe the 'homogeneity and isotropy problem' and explain how the idea of inflation offers one solution to it.

8 Explain how the idea of inflation offers a possible solution to the problem of why the density of the Universe is close to the critical value.

9 Give a brief outline history of the Universe from 10^{-43} s to the present day.

Study comment

As mentioned at the start of Unit 14, whereas that was a relatively light Unit, this one has a work load somewhat greater than average. So, you should make an early start on Unit 15. The one that follows, Unit 16, is a revision unit, so it contains no extra subject matter, but you are advised not to use that as a reason for allowing the work of Unit 15 to encroach on that revision period.

Strategy if you get behind. If you are getting behind, the cleanest option is to omit Section 6.

---

1   Introduction

We now use general relativity to examine the structure and evolution of the Universe.

Band 5 of AC4 introduces this Unit.

In Block 3 we saw how an aggregate of matter, such as the Sun, was able to distort spacetime in its vicinity. Now we shall discover that the average energy density throughout the Universe produces an *overall* curvature. This curvature determines whether the Universe is infinite in extent (an open universe), or alternatively occupies a finite volume — but again without a boundary (a closed universe). It also provides the key to predicting what the future holds for the Universe: indefinite expansion, or an expansion that is one day halted, leading thereafter to contraction and a big crunch. Thus, the determination of the energy density assumes crucial importance. Unfortunately, it appears that the bulk of the matter in the Universe is not visible. That which we can see in the form of visible stars represents only a small fraction of the whole. The rest is contained in what we call 'dark' matter. One of the great problems we shall face is that of determining how much dark matter there is, and what its composition might be.

We shall discover that the average energy density sits very close to the critical value that corresponds to a flat spatial geometry $(k = 0)$. This is such a strange state of affairs that it seems to demand an explanation of some sort. It is in this context that it has been postulated that at a very early stage in the development of the Universe — within a tiny fraction of a second of the instant of the big bang, there was an exceedingly brief, but crucial period of superfast expansion called 'inflation'. This is a process that automatically produces a universe in which the spatial geometry is effectively flat $(k/R^2 = 0)$. Because the subjects of dark matter and inflation are so important to determining the future of the Universe, they will figure prominently in our discussions. Another speculative topic that will arise is that of dark energy. This concerns the possibility that some of the cosmic energy density, possibly even the majority, may not be attributable to any form of matter (dark or bright) nor even to radiation, but rather to some entirely different source.

Finally, armed with all the information gleaned from this Block, we venture to outline, to the best of our knowledge, how the Universe has evolved since a fraction of a second after the instant of its formation, to the present day, and how it is likely to go in the future.

2 The Friedmann–Robertson–Walker cosmology

2.1 The effect of matter and energy on spacetime

As we said in Block 3, the content of the general theory of relativity has been summarized by the noted physicist John Wheeler, in the statement: 'curved spacetime tells matter how to move, and matter tells spacetime how to curve'. So far, we have tended to concern ourselves mostly with the first half of this statement. We have assumed a certain metric, such as the flat metric of special relativity, and concentrated on *using* it to explain the motion of particles. In the previous Unit we continued this emphasis by quoting the form of the Robertson–Walker metric and using it to explain the observations of Hubble and others on the redshifts of galaxies. This approach culminated in Equation 11, Unit 14, which gives a direct link between the observed redshift z, and the metric parameter $R(t)$.

But, in order to make further progress, we must now consider the other facet of general relativity; the way in which a given distribution of matter or energy causes spacetime to curve. In general relativity, there is a well-defined procedure for finding the metric that is appropriate for a given situation — one must solve Einstein's field equations. In Unit 10 & 11, we saw this in connection with the Schwarzschild metric produced outside a spherically symmetric distribution of matter. In this Unit we shall use Einstein's field equations to answer questions of cosmological significance. But first, we shall pause to recall the meaning of these equations and the role they play in general relativity as a whole.

In their most general form, Einstein's field equations are very complicated. But, as we saw in Section 12.7 of Unit 10 & 11, they can be written schematically in the form:

$$\boxed{\begin{array}{l}\text{Ricci curvature of} \\ \text{spacetime at any} \\ \text{given event}\end{array}} = \boxed{\begin{array}{l}\text{Ten-component source} \\ \text{term at the same event}\end{array}}.$$

The solution gives us the metric coefficients everywhere from which the curvature of spacetime can be calculated. In Unit 10 & 11 we spent a long time clarifying the meaning of curvature with the aid of two-dimensional analogies. We saw that, in two dimensions, the curvature at a given point is described by a single number, $\mathscr{K}$, which can be calculated directly from the metric.

In fact, if the metric is

$$(\Delta l)^2 = g_1(\Delta q^1)^2 + g_2(q^1)(\Delta q^2)^2$$

where g_1 and g_2 are the metric coefficients at the point (q^1, q^2), it follows that the curvature $\mathscr{K}(q^1, q^2)$ at the point (q^1, q^2) is given by

$$\mathscr{K}(q^1, q^2) = \frac{1}{4g_1[g_2(q^1)]^2}\left(\frac{\mathrm{d}g_2}{\mathrm{d}q^1}\right)^2 - \frac{1}{2g_1 g_2(q^1)}\left(\frac{\mathrm{d}^2 g_2}{\mathrm{d}q^{12}}\right).$$

In four-dimensional spacetime, the curvature takes a far more complicated form; in fact *twenty* numbers are needed to determine the Riemann curvature at any given event. But each of these 20 numbers can still be expressed in terms of metric coefficients and their first and second derivatives with respect to the coordinates. So, although the mathematics becomes more cumbersome, the concepts remain the same.

On the right-hand side of the equation are the agents which *cause* the curvature of spacetime. As we pointed out in Unit 10 & 11, *three* different types of quantity are involved. They are:

1 the energy per unit volume (which we denote by the symbol ρ);

2 the momentum per unit volume;

3 the momentum flux.

There are several points we would like you to understand about these source terms.

Firstly, it is important to realize that both energy and momentum are to be understood in their *relativistic* sense: for example, the contribution to the total energy due to a free particle with speed v is

$$E = cp^0 = cm\frac{\Delta x^0}{\Delta \tau} = \frac{mc^2}{\sqrt{1 - \dfrac{v^2}{c^2}}}. \qquad \text{(Unit 7, Eq. 41)}$$

As shown in Equation 34 of Unit 7, this energy contains *both* matter energy (mc^2) and kinetic energy ($\frac{1}{2}mv^2 +$ higher-order terms). In order to remind ourselves of this fact, we shall refer to ρ as the *energy density*, expressed as an energy per unit volume, in units $\mathrm{J\,m^{-3}}$. Sometimes we shall refer to the *mass density*; this is given by ρ/c^2, and is expressed in units of $\mathrm{kg\,m^{-3}}$.

energy density

mass density

Secondly, we must be careful to include *all* contributions to the energy and momentum. You will remember from Unit 4 that fields carry energy and momentum just as surely as particles do. We therefore need to count the relativistic energies of the particles *together with* the energies of any fields that happen to be present. But, as far as Einstein's field equations are concerned, the origin of the energy or momentum is irrelevant. This is rather important for cosmology, because the nature of the Universe has changed dramatically during its lifetime. At the present time the energy and momentum of *galaxies* are relevant. Earlier, it was *radiation* that dominated both quantities. In the relativistic formulation, the same equations apply to both phases of the Universe.

Finally a word about the momentum flux. As its name implies, momentum flux describes the flow of momentum from one region of space to another. For example, the gas of particles in Figure 1 has a large **1**-component of momentum in region A, and a small **1**-component of momentum in region B.

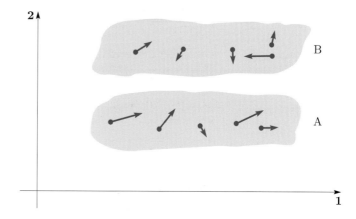

Figure 1 The movement of particles in a gas, in two regions A and B.

As time passes, the motion of the gas particles will cause this distribution of momenta to change: because the gas particles in region A also have a significant component in the **2**-direction towards region B, *momentum will flow from region A to region B*. The amount of momentum transported across unit area in unit time is called the momentum flux.

In general, one needs to worry about *which* component of momentum is being transported in *which* direction — in the above example, we were concerned with the **1**-component of momentum being transported in the **2**-direction. But, in cosmology, a remarkable simplification becomes possible. According to the cosmological principle, the distribution of energy and momentum in our Universe is, and always has been, isotropic and homogeneous. It follows that, at any given time or place, the mean momentum density (averaged over a region large enough to contain many clusters) must be zero. Otherwise, the momentum vector would select out a special direction in space — and that is just what the cosmological principle denies.

At first sight, the cosmological principle also seems to imply that the momentum flux is zero — but that would be to forget about the expansion of space! Hubble's observations of redshifts have shown us that the whole Universe is expanding, so some momentum must be carried outwards from every point. Thus, the momentum flux in the Universe may not be zero but is characterized by a single value at any time. Just one value is involved, because space is expanding equally fast in all directions. This value is generally written as p, and is referred to as the *pressure*. (The nomenclature turns out to be quite helpful because p plays a role similar to that of the pressure in an ordinary gas.)

pressure

Finally, because of the homogeneity of space, both the mass density and the pressure must be uniform throughout all space. We therefore arrive at a rather surprising conclusion:

> For the purposes of cosmology, the contents of our Universe can be accurately modelled by just *two* functions of time, $\rho(t)$ and $p(t)$.

The immense complexity of the fine-scale details can be effectively ignored.

Objective 1   **SAQ 1**   Different possible homogeneous universes are characterized mainly by their densities and pressures, as functions of time. In our idealized model there are not many other parameters that can vary. However, another possibility is composition. Our Universe consists mostly of hydrogen. Would a gold universe of the same density and pressure behave differently, from the cosmologist's point of view? What about antihydrogen or even antigold?

2.2   The Friedmann equations

We saw in Unit 14 how acceptance of the cosmological principle for a Universe undergoing Hubble expansion, constrains the metric to be of the form proposed by Robertson and Walker. But Einstein's field equations tell us more than this. They give a precise relationship between the two parameters $R(t)$ and k that occur in the metric, and the two quantities $\rho(t)$ and $p(t)$ that describe the contents of our Universe.

Friedmann equations

In fact, applying Einstein's field equations to a universe described by the Robertson–Walker metric, we find the so-called *Friedmann equations*

$$\frac{1}{R(t)}\frac{\mathrm{d}^2 R(t)}{\mathrm{d}t^2} = -\frac{4\pi G}{3c^2}(\rho(t) + 3p(t)) \tag{1}$$

$$\frac{1}{R(t)}\frac{\mathrm{d}^2 R(t)}{\mathrm{d}t^2} + \left(\frac{1}{R(t)}\frac{\mathrm{d}R(t)}{\mathrm{d}t}\right)^2 + \frac{kc^2}{R^2(t)} = \frac{4\pi G}{3c^2}(\rho(t) - 3p(t)). \tag{2}$$

cosmological constant

Equations similar to these were first put forward by A. Friedmann in 1922 and later, but independently, by G. Lemaitre. In fact they considered a slightly different case with an additive constant on the right-hand side (as introduced by Einstein) called the *cosmological constant*.

Einstein introduced the cosmological constant for reasons that are now recognized as largely bogus. He is alleged to have described its introduction as his 'greatest blunder'. However, the cosmological constant, usually denoted Λ, has had a controversial history in cosmology and is very much back in vogue at present. Its most recent renaissance is partly (though not entirely) a consequence of the observational indications that cosmic expansion is accelerating. The cosmological constant, as introduced by Einstein, has the effect of producing a large-scale repulsion that could account for the observations of acceleration.

Although we wish to take account of the cosmological constant in some of the discussions that follow, we do not plan to reintroduce Λ into the Friedmann equations. Rather, we adopt an approach widely favoured by modern cosmologists, which is to recognize that all the effects of the cosmological constant can be reproduced by assuming that one of the contributions to the total energy density ρ may come from what is termed *dark energy* . This dark-energy contribution is usually denoted ρ_Λ. In order to mimic the cosmological constant it must behave rather oddly. In particular, its value must remain constant, despite the expansion of the Universe. Also, associated with the dark-energy density, there must be a contribution to the pressure p, and this contribution, denoted p_Λ, must have the value $p_\Lambda = -\rho_\Lambda$. In the context of general relativity, this negative cosmic pressure can be thought of as promoting the accelerating expansion that recent observations indicate.

dark energy

Although you are not expected to remember the exact form of Equations 1 and 2, you need to be able to recognize the significance of their various features. We begin by noting that the right-hand sides contain the physical quantities that determine the curvature of spacetime: ρ, the average energy density in the Universe, and p, the pressure. Neither p nor ρ vary in space (a result of the cosmological principle), but they do vary throughout the life of the Universe. The terms that appear on the left-hand sides of

Equations 1 and 2 will also be familiar to you:

$$\frac{1}{R}\frac{dR}{dt} \quad \text{is the Hubble parameter, } H \qquad \text{(Unit 14, Eq. 13)}$$

$$\frac{1}{R}\frac{d^2R}{dt^2} \quad \text{is the deceleration parameter, } q, \qquad \text{(Unit 14, Eq. 14)}$$
$$\text{multiplied by } -H^2$$

$$\frac{k}{R^2} \quad \text{is the curvature of space} \qquad \text{(see Unit 14, Section 4.4)}$$

Study comment

In order to simplify the nomenclature, we have here written $R(t)$ as R. From now on, we shall in general take the time dependence of the quantities R, H, q, p and ρ as implicit.

Objective 1   **SAQ 2**   Rewrite the Friedmann equations in this somewhat simplified notation.

Objective 1   **SAQ 3**   Is it also implicit that the parameter k appearing in the Robertson–Walker metric and in the Friedmann equations is dependent on t?

In general, the curvature of spacetime could depend on the metric coefficients and their first and second derivatives with respect to the coordinates. But, in a homogeneous universe, only variations in time are important, so Equations 1 and 2 contain only k, R, dR/dt and d^2R/dt^2.

If we knew the values of $\dfrac{1}{R}\dfrac{dR}{dt}$, $\dfrac{1}{R}\dfrac{d^2R}{dt^2}$ and $\dfrac{k}{R^2}$ for all t, we would know the way in which the geometry of spacetime evolves in time. We would then be able to answer many fundamental questions about the age and future of our Universe. Unfortunately, as you know, there are severe practical problems in measuring quantities like the deceleration parameter, and hence $\dfrac{1}{R}\dfrac{d^2R}{dt^2}$. We are therefore forced, from now on, to adopt a less direct strategy, which may be summarized as follows:

1 We first obtain estimates for the present values of ρ and p.

2 We then make reasonable assumptions about the way in which ρ and p depend upon R.

3 Finally, we use the Friedmann equations to deduce R, based on assumed values of k.

Much of Section 4 will be concerned with the first step in this plan. For example, we shall discuss different methods for estimating the present energy density, ρ_0. It is necessary to study the experimental evidence rather closely because the value of ρ_0 turns out to have a crucial influence on the future evolution of the Universe.

You are reminded that the suffix $_0$ refers to the value of a quantity evaluated at the present time, $t = t_0$.

The second step is to find how ρ has varied with R. This is an even more difficult task than finding ρ_0 and observations are of little direct help. We

dark-energy-dominance

matter-dominance

radiation-dominance

must therefore use a model. There are three possible phases of the evolution of the Universe which we can clearly distinguish, and we shall ignore the complicated processes during the transition time between them. Nowadays we may be in a phase which is *dominated by dark energy* as a result of which the energy density is effectively constant, independent of R. Prior to that we are believed to have been in a phase *dominated by the energy density due to various forms of matter* when the energy density was proportional to $1/R^3$. In an even earlier phase, however, the Universe was *dominated by radiation*, the energy density of which was proportional to $1/R^4$. These three phases are indicated in Figure 2.

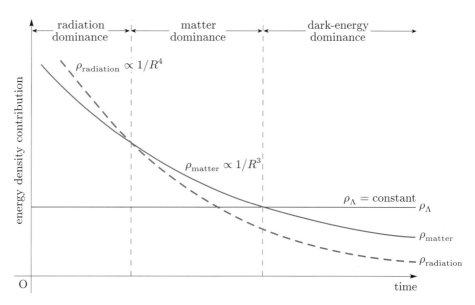

Figure 2   A schematic illustration of the variation with time of three plausible contributions to the average energy density of the Universe; radiation ($\rho_{\text{radiation}} \propto 1/R^4$), matter ($\rho_{\text{matter}} \propto 1/R^3$) and dark energy ($\rho_\Lambda = \text{constant}$). Each of these contributions may be dominant at some stage in the evolution of the Universe.

energy density of dark energy

The fact that dark-energy density is independent of R follows from its definition. It is also easy to see why radiation and matter should behave in distinct ways. Remember that R is a scale factor which multiplies comoving coordinate differences in the Robertson–Walker metric by the same factor, converting them to lengths. Since volume has the dimension of $[\text{length}]^3$, it is clear that the size of a given volume in space (defined with the aid of chosen galaxy clusters and geodesics) is proportional to R^3. We shall assume that the total energy content of matter is unaffected by expansion. It therefore follows that the energy *per unit volume* due to matter is

energy density of matter

$$\rho_{\text{matter}} = \frac{\text{constant}}{\text{volume}} \propto \frac{1}{R^3}. \tag{3}$$

At first sight, the same conclusion also seems to apply to radiation. But the situation here is slightly more complicated. The energy of a photon is proportional to its frequency and hence inversely proportional to its wavelength:

$$\begin{aligned} E &= hf \\ &= h\frac{c}{\lambda}. \end{aligned} \tag{Unit 13, Eq. 8}$$

In the case of radiation we must allow for the fact that wavelengths are affected by the expansion of space, so that

$$\lambda \propto R.$$

Thus

$$E \propto \frac{1}{\lambda} \propto \frac{1}{R}.$$

This means that, as the Universe grows larger, individual photons become *less* energetic. This introduces an extra factor of $1/R$ into the expression for ρ; so for radiation,

$$\rho_{\text{radiation}} \propto \frac{1/R}{\text{volume}} \propto \frac{1}{R} \times \frac{1}{R^3} = \frac{1}{R^4}. \tag{4}$$

energy density of radiation

In the next section we shall combine these models for the behaviour of ρ with the third step in our plan: applying Equations 1 and 2 to questions of cosmological importance. It is sensible to begin here because we shall then see how important it is to gain an accurate estimate of today's energy density, ρ_0 — the subject of Section 5.

Study comment

For our purposes, it is desirable to simplify matters by eliminating p between Equations 1 and 2. This is done in SAQ 4. Because we shall be using the resulting combined equation quite frequently in the next Section you are advised *not* to omit this SAQ, or SAQs 5 and 6.

Objective 1   **SAQ 4**   Show that Equations 1 and 2 can be combined to give

$$\left(\frac{\mathrm{d}R}{\mathrm{d}t}\right)^2 + kc^2 = \frac{8\pi G}{3c^2}\rho R^2. \tag{5}$$

Use this equation to comment on the value of k if the Hubble parameter were zero.

Objective 1   **SAQ 5**   Show from Equation 1 that, under the condition that p is negligible, the rate of expansion of the Universe must be decreasing. (This corresponds to an epoch of matter-dominance.)

Objective 1   **SAQ 6**   Use Equation 5 to find the particular value of the energy density today, ρ_0, for which $k = 0$. Write your result in terms of the Hubble parameter evaluated today, H_0.

Summary of Section 2

1 The agents that cause the curvature of spacetime are (i) the energy density, ρ; (ii) the momentum density; and (iii) the momentum flux.

2 Because of the homogeneity of space, the value of the first contribution, the energy density ρ, is the same throughout space.

3 Because of the isotropy and homogeneity of the Universe, the second contribution, the momentum density, is zero.

4 Because of the uniform expansion of the Universe, the third contribution, the momentum flux, is described by a single quantity, p, referred to as the pressure.

5 Applying Einstein's field equations to a universe described by the Robertson–Walker metric yields the Friedmann equations (Equations 1 and 2). These link the quantities R and k of the metric to the two non-zero agents producing curvature, ρ and p.

6 The energy density, ρ, is made up of a contribution due to radiation, $\rho_{\text{radiation}}$, a contribution due to matter, ρ_{matter}, and (perhaps) a contribution due to dark energy, ρ_Λ. Because the density dependences of these three are not the same ($1/R^4, 1/R^3$ and constant respectively) the Universe has gone from a period of radiation-dominance in the early stages, to one of matter-dominance later on, and may now be in a period of dark-energy-dominance.

3 The Friedmann equations and the evolution of the Universe

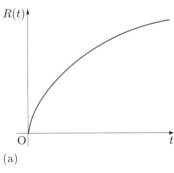

(a)

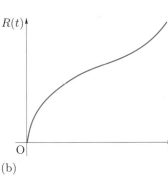

(b)

Figure 3 (a) Variation of R with t throughout the eras of radiation-dominance and matter-dominance. R is equal to zero at some point in time called $t = 0$. (b) Variation of R with t including a recent period of dark-energy-dominance.

* Because ρ_Λ is much lower than [?] Predictive or Power of the [?] are $\propto \frac{1}{R^4}$ & $\frac{1}{R^3}$ respectively

The fact that the galaxies have redshifts, not blueshifts, means that the present value of dR/dt is positive, not negative (i.e. the distance between galaxy clusters is increasing). Thus a graph of R against time must be increasing with t, at the present time t_0.

So much for the *direct* information about R that is revealed by current astronomical observations. But Friedmann's equations allow us to progress much further. For example, as was shown in SAQ 5, if p is negligible

$$\frac{d^2R}{dt^2} = -\frac{4\pi G}{3c^2}(\rho + 3p)R$$
$$= -\frac{4\pi G}{3c^2}\rho R$$
$$< 0.$$

This indicates a slowing down of cosmic expansion, which is probably what happened throughout the periods of radiation-dominance and matter-dominance. Thus, ignoring for the moment any (recent) era of dark-energy-dominance, the Friedmann equations imply that the scale factor varied with time in the way shown in Figure 3(a).

Notice that even if p were not assumed to be negligible, as was done in SAQ 5, the effect of a positive pressure p would be to cause an added *deceleration*. This can be traced to the fact that the momentum flux acts as a source term in the Einstein field equations in the same way that mass does. Even a negative pressure, equal to $-\rho_\Lambda$, would have had a negligible effect throughout the eras of matter-dominance and radiation-dominance.* So over most of the history of the Universe, the effect of *both* density and pressure has been to cause dR/dt to decrease with time. This is the general relativistic version of the idea that attractive gravitational forces must slow down the rate of expansion. Thus although the graph of R against t must be increasing with t it must also be curving downwards, as in Figure 3(a) (i.e. it is *concave downwards*), throughout most of cosmic history. In a similar way, if we accept the current evidence for accelerating cosmic expansion and associate it with a dominant dark-energy density ρ_Λ and a negative pressure $p = -\rho_\Lambda$, then Equation 1 implies

$$\frac{d^2R}{dt^2} = -\frac{4\pi G}{3c^2}(\rho + 3p)R$$
$$= \frac{8\pi G}{3c^2}\rho_\Lambda R$$
$$> 0.$$

In this case, the Friedmann equations imply that the graph of R against t should curve upwards during the era of dark-energy-dominance, giving the overall result shown in Figure 3(b) (i.e. it becomes *concave upwards* after sufficiently large t).

3.1 Looking into the past

Let us consider the values that R has taken in the past. Since R decreases more and more rapidly the further back one goes in time, the graph in Figure 3(b) must cut the t-axis somewhere.

> In other words, the Friedmann equations say that there must have been a time when $R = 0$, that is, when the Universe was infinitely dense. The sudden expansion from this initial state at $t = 0$ is what we have been calling the big bang.

If we wish to know how long ago the big bang occurred, we need to calculate the exact shape of the curve of R against t; and that requires extra assumptions about the way ρ and p have depended on time.

age of the Universe   Quantitative estimates of the age of the Universe will be discussed later in this Unit.

3.2   Looking into the future

Friedmann's equations can be used not only as a guide in looking backwards but they also enable us to look forwards. We are now poised to do this, having defined our terms, including $t = 0$, and having set out Equations 1 and 2. Looking forward will underline the importance of knowing the energy density of the Universe, which is a major concern of this Unit.

We begin by considering what will happen in the distant future, as t becomes very large, if a constant dark-energy density really is dominant. In this case it follows from Equation 5 that

$$\frac{1}{R^2}\left(\frac{\mathrm{d}R}{\mathrm{d}t}\right)^2 + \frac{kc^2}{R^2} = \frac{8\pi G\rho_\Lambda}{3c^2}.$$

As the accelerating expansion increases both R and $\mathrm{d}R/\mathrm{d}t$, the term kc^2/R^2 will eventually become negligible. Ignoring this term and taking the square root of both sides of the remaining equation gives

$$\frac{1}{R}\frac{\mathrm{d}R}{\mathrm{d}t} = \left(\frac{8\pi G\rho_\Lambda}{3c^2}\right)^{1/2}.$$

It is easy to verify (by substitution if desired) that this first-order differential equation has the solution

$$R(t) = A\exp\left[\left(\frac{8\pi G\rho_\Lambda}{3c^2}\right)^{1/2}t\right]$$

where A is an arbitrary constant. Thus if our Universe indeed turns out to be dominated by dark energy, its future is clear. Clusters of galaxies will separate at an exponentially increasing rate. The distances between them will increase by a factor of e in each epoch of order 10^{10} years. Yet the structure of own Galaxy will be unaffected by this ever accelerating cosmic expansion, since the local energy density of matter within the Milky Way is – and will remain – far greater than any supposed dark-energy density. Eventually, the only light signals that will reach the Milky Way will come from other galaxies in our Local Group, or Supercluster, which will become oblivious to the rest of the Universe.

It is only recently that such a dark-energy scenario has been widely entertained. From 1922 to the end of the last century, it was generally believed that the future of the Universe would be determined by the matter that it contained. Moreover it was not known whether the amount of matter was sufficient to halt the expansion of the Universe. In the remainder of Section 3 and for much of Section 4 we shall be applying the

Friedmann equations to a matter-dominated universe. Then we shall return to the issue of dark energy at the end of Section 5. Please note that SAQs 7 to 20 make no reference to dark energy and are to be answered on the hypothesis that there is no such constant contribution to the large-scale energy density of the Universe. In this development we shall thus be retracing key arguments that preoccupied the majority of cosmologists until very recently.

So let us now agree to ignore the evidence for dark-energy-dominance and consider the fate of a universe presently dominated by matter in which $\rho_\Lambda = 0$, like the one shown in Figure 3(a). Recall from SAQ 5 that in such a universe $\mathrm{d}^2 R/\mathrm{d}t^2$ is always negative. It follows that $\mathrm{d}R/\mathrm{d}t$ gets smaller as time increases, and that is why the graph of R against time bends over as in Figure 3(a). This raises a question about the future of a matter-dominated universe: does R reach a maximum value and then decrease back to zero, or does it increase forever but at an ever slower rate?

closed universe

In the former case, the universe is said to be *closed*. If our Universe were closed, there would in the future be a contracting phase (for $t > t_\mathrm{m}$ in Figure 4), when R would decrease and light from the galaxies would be blueshifted. One may speculate whether there would also be fundamental effects on the laws of physics. Certainly the very final phases of contraction, like the very early phases of the big bang, would be governed by physical laws that are as yet unknown. We call this speculative

big crunch

contraction a *big crunch*. What would happen *after* the Universe had contracted we have no way of telling. One suggestion is that the Universe would go out of existence at the instant of the big crunch. Another is that it would begin to expand once more. This latter scenario we call a *big*

big bounce

bounce; the Universe would oscillate, alternately expanding and contracting. That being the case, what we have been calling the big bang would merely be the most recent of the big bounces.

The other possibility for the future variation of R in a matter-dominated universe is that R has no maximum value and the universe continues expanding forever (Figure 5). A universe which behaves in this way is said

open universe

to be *open*.

The Friedmann equations make two specific predictions about open and closed universes. The first is concerned with geometry, and the second with mass density. Both predictions can be derived from Equation 5:

$$\left(\frac{\mathrm{d}R}{\mathrm{d}t}\right)^2 = \frac{8\pi G}{3c^2}\rho R^2 - kc^2.$$

It is clear that if R reaches a maximum value at, say, $t = t_\mathrm{m}$, (see Figure 4) then

$$\left(\frac{\mathrm{d}R}{\mathrm{d}t}\right)_{t=t_\mathrm{m}} = 0$$

at that maximum, so that we know

$$k = \frac{8\pi G}{3c^4}\rho(t_\mathrm{m})R^2(t_\mathrm{m}) > 0.$$

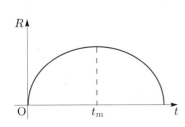

Figure 4   Behaviour of R in a closed universe which corresponds to positive curvature ($k = +1$).

We conclude that there is a link between the geometry of space and the future behaviour of the Universe. If the Universe is *closed*, k must be $+1$ and space must have a *positive* curvature — meaning, for example, that the angles in a triangle (drawn with a supercluster at each vertex) would sum to *more* than 180°. Space in all its regions would have the same value of positive curvature, k/R^2, at time t in a closed Universe, in much the

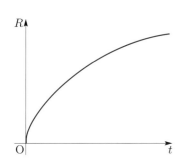

Figure 5 Behaviour of R in an open universe, which corresponds to negative curvature ($k = -1$).

same way that the two-dimensional surface of a sphere of radius a, is described by a single value of positive curvature, $1/a^2$. Thus, if our Universe is *closed*, it would have a *finite* (but expanding) volume of space, in the same way that the surface of an inflating balloon has a *finite* (but expanding) surface area. Indeed the choice of the word 'closed' to describe a universe with $k = +1$ indicates the possibility that three-dimensional space could 'close round' on itself, in the same way that the two-dimensional surface of a sphere does. Hence a closed universe could be finite in volume (and hence in mass) without having any boundaries. We cannot produce a picture or model with these properties, because the closure of three-dimensional space could only be represented in a fourth spatial dimension. The best *analogy* for a closed Universe is the expanding balloon: you can buy a model of a closed Universe with only *two* spatial dimensions in any three-dimensional toy shop.

On the other hand, if R has no maximum, i.e. $\left(\dfrac{dR}{dt}\right)^2$ is never zero (see Figure 5), the Universe is, as we said, open. Under these circumstances

$$\left(\frac{dR}{dt}\right)^2 > 0$$

so

$$\frac{8\pi G}{3c^2}\rho R^2 - kc^2 > 0.$$

Because ρ diminishes as $1/R^3$ if matter is dominant, or as $1/R^4$ if radiation is dominant, the first term on the left-hand side approaches zero with time. Thus, for large values of t, the inequality can be satisfied only if $k = 0$ or -1. An *open* Universe is therefore characterized by a *zero or negative* curvature of space, with the angles in a large triangle summing to 180° or *less*. In an *open* Universe, space *cannot* 'close round' on itself in any simple way, for the same sort of reason that you cannot make a smooth horn or saddle (negative curvature) or plane (zero curvature) of finite area, with no edges. So presumably, the space of an *open* ($k = 0$ or -1) Universe is *infinite*, extending in all directions, with no boundaries. If we inhabit an open Universe, we just happen to be in one particular region of an infinite space, much like any other region. As the Universe expands we shall receive light signals from more and more of the infinite number of galaxies it contains. Some people might find the prospect of an infinite universe unpalatable, preferring to contemplate a closed Universe of finite size and finite mass. But of course, the true fate of the Universe will be determined by observation, not by personal preference.

In principle, it is possible to predict whether the Universe will eventually contract by measuring k in observations like those of SAQ 5 in Unit 14. But until we can communicate with beings in other galaxy clusters we must use other methods! One of these is based on the density of mass.

It is apparent that the future of a matter-dominated universe depends on the sign of the parameter k in the metric. Recalling that

$$H \equiv \frac{1}{R}\frac{dR}{dt}$$

we can see from Equation 5 that k and R are related to observable parameters by

$$\frac{kc^2}{R^2} = \frac{8\pi G}{3c^2}\rho - H^2. \tag{6}$$

If $k = +1$, the right-hand side must be positive, implying that

$$\frac{8\pi G}{3c^2}\rho > H^2.$$

Thus

$$\rho > \frac{3c^2}{8\pi G}H^2.$$

In the limiting case, the density is given by the equality

$$\rho^c = \frac{3c^2}{8\pi G}H^2. \tag{7}$$

critical density, ρ^c

But this is similar to the expression for the density we derived for the $k = 0$ case in SAQ 6. The energy density appearing in Equation 7 is given a special name — the *critical density*. It is generally denoted by the symbol ρ^c and is a function of time, ρ_0^c referring to its value at the present time, t_0. By substituting the present value of the Hubble parameter $(H_0 \approx 2.3 \times 10^{-18}\,\text{s}^{-1})$ in this equation, we can find out how large ρ_0^c is:

From Unit 13 we have $H_0 = h \times 100\,\text{km}\,\text{s}^{-1}\,\text{Mpc}^{-1}$. With a current estimate of $h = 0.7$, this yields $H_0 \approx 2.3 \times 10^{-18}\,\text{s}^{-1}$.

$$\rho_0^c = \frac{3c^2}{8\pi G}H_0^2$$

$$= \frac{3 \times (3 \times 10^8)^2 \times (2.3 \times 10^{-18})^2}{8 \times \pi \times (6.67 \times 10^{-11})}\,\text{J}\,\text{m}^{-3}$$

$$= 8.5 \times 10^{-10}\,\text{J}\,\text{m}^{-3}.$$

It is common to see ρ_0^c expressed as the mass density $\rho_0^c/c^2 \approx 10^{-26}\,\text{kg}\,\text{m}^{-3}$. Much of the rest of this Unit is concerned with estimating the actual value of ρ_0 and checking whether it is greater or smaller than ρ_0^c.

Three key points have now emerged:

1 The Universe is expanding from an initial high-density condition, called the 'big bang', which existed a finite time ago.

2 The long-term future of our Universe, if devoid of dark energy, depends on its present density, ρ_0, and the present value of the Hubble parameter, H_0.

3 The long-term future of our Universe, if dominated by dark energy, is in exponentially growing expansion, leading to the isolation of our Local Group from the rest of the Universe.

Objective 2   **SAQ 7**   Would $\text{d}^2R/\text{d}t^2$ be negative in a contracting phase?

Objective 2   **SAQ 8**   Ignoring any dark energy, sketch very roughly the variation of density with time, assuming (a) a closed universe with $k = +1$, and (b) an open universe with $k = -1$. What is the lowest value of ρ that can be reached in each case, expressed in terms of the maximum value of R?

Objective 2   **SAQ 9**   Suggest some of the differences in the night sky that would be seen by an observer on Earth if the Universe were closed and had passed the peak in R.

4   The age of the Universe

We have often referred to the present time as t_0, implying that t_0 is the age of the Universe. But so far, no indication has been given of how the value of t_0 might be deduced from observations. That is the issue we shall now address. We shall start by considering a universe devoid of dark energy and finding an upper limit on its age; only then will we consider the more general case of a universe that may contain dark energy.

4.1   Upper and lower limits to the age

We have seen that in a universe containing only matter and radiation (*not* dark energy), the scale factor increases with time at a decelerating rate, as shown in Figure 3(a). In such a universe where $\mathrm{d}^2R/\mathrm{d}t^2 < 0$, an upper limit to the age of the universe can be obtained by drawing a tangent to the curve relating R and t, touching the curve at the time corresponding to the present t_0 (see Figure 6). Extrapolating this tangent back to $R = 0$ indicates a time t_{ex} (for 't extrapolated'). Because $\mathrm{d}^2R/\mathrm{d}t^2 < 0$, the true value of R at any time t will always be below the tangent line at that value of t. Hence $(t_0 - t_{\mathrm{ex}})$ is an upper limit for the age of this matter-dominated universe.

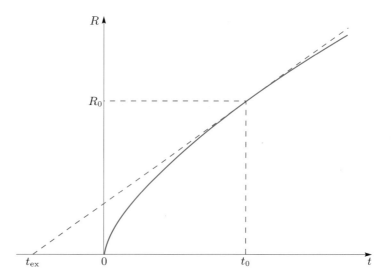

Figure 6   How to get an upper limit for the age of the Universe from a plot of R against t.

Clearly from the figure, the gradient of the tangent $(\mathrm{d}R/\mathrm{d}t)_{t=t_0}$ is

$$\left(\frac{\mathrm{d}R}{\mathrm{d}t}\right)_{t=t_0} = \frac{R_0}{t_0 - t_{\mathrm{ex}}}.$$

But $\dfrac{1}{R_0}\left(\dfrac{\mathrm{d}R}{\mathrm{d}t}\right)_{t=t_0}$ is the current value of the Hubble parameter, H_0.

Thus

$$t_0 - t_{\mathrm{ex}} = \frac{1}{H_0}$$

and since t_{ex} is negative, this means

$$t_0 < \frac{1}{H_0}. \tag{8}$$

Objective 3   **SAQ 10**   Accepting the value $H_0 = 2.3 \times 10^{-18}\,\text{s}^{-1}$ given in Unit 13, evaluate the upper limit on t_0 in seconds and in years.

Historically, an upper limit obtained in this way led to one of the first tests of cosmological ideas, because it could be compared with a completely independent estimate of the age of matter using nuclei that decay slowly. Such methods are called radioactive-dating methods and they are used in a number of fields. The possibility of comparing estimates of the age of the Universe based on two totally different approaches was of tremendous significance to the study of cosmology.

Since the time of those early comparisons, there has been a great improvement in the determination of H_0. Also a number of other ways have been devised of estimating the age of the Universe independently of cosmological considerations. These developments make the comparison even more interesting and potentially more informative.

Here are three methods of estimating the age of ancient objects in our Universe. Each of these estimates provides a *lower limit on the age of our Universe* that can be compared with the upper limit provided by the current value of the Hubble parameter.

1 *Nuclear dating*   As already indicated, this method is based on slowly decaying nuclei. In practice the nuclei that are used have mean lifetimes comparable to the age of the Universe and the dating technique depends on measuring the ratios of pairs of long-lived nuclei in a variety of ancient samples. Based on this technique, there is good agreement that the age of the Moon is about 4.6×10^9 years, and that the Universe was probably at least 5×10^9 years old by the time the Moon — and the rest of the Solar System with it — was formed. So this method of estimating a *lower* limit to the age of the Universe gives a result consistent with the *upper* limit for a matter-dominated universe of about 1.4×10^{10} years obtainable from the Hubble parameter.

lower limit for age of Universe

2 *The age of the oldest star clusters*   A more stringent lower limit on the age of the Universe is obtained by studying the stars belonging to *globular clusters*. These clusters consist of up to a million stars each. They are situated mainly outside the plane of the disk of the Galaxy, and this suggests that they probably originated in the very earliest stages of the formation of the Galaxy — before it had settled down to its characteristic flat-disk shape. So, the stars of globular clusters are old. By examining a plot of their brightness versus colour (the so-called Hertzsprung–Russell diagram) it is possible to estimate the age of such stars. Widely accepted theories of stellar development place the age within the range $(0.85$ to $1.33) \times 10^{10}$ years with 1.2×10^{10} years as a likely value. Because the stars could not have formed before the Universe came into existence, this gives us a lower limit on the age of the Universe — in fact our best estimate of that lower limit. This latter result lies within the upper limit gained from the Hubble parameter.

3 *The age of the oldest white dwarf stars*   White dwarf stars are highly evolved stars of the kind that the Sun will become when it has consumed all the nuclear fuel it currently contains. The luminosity of a white dwarf results from the energy it retains after nuclear reactions have ceased in its interior. Hence each white dwarf is gradually fading as its internal energy leaks out and the faintest white dwarfs are generally the most ancient.

Studies of faint white dwarfs indicate ages of 1.2×10^{10} years with an uncertainty of about 0.1×10^{10} years. Again this is consistent with the upper limit from the Hubble parameter.

So far, the results of comparing the lower age limits set by various ancient objects with the upper age limit provided by cosmology have been encouraging. But remember, it is an upper limit that we have been comparing with, not an actual age. Furthermore, we have only been considering a matter-dominated universe; we have not yet given any consideration to the possible effect of dark energy. We will attend to both these issues in the next section.

Objective 3   **SAQ 11**   Draw a time-scale, on which you indicate known lower limits for the age of the Universe, labelling each limit appropriately. Also indicate the upper limit on the age of a matter-dominated universe indicated by the current value of the Hubble parameter.

4.2   *Estimate of the age of the Universe*

To get an estimate of the age of a matter-dominated universe, rather than an upper limit, one needs not only a good measurement of H_0, but also knowledge about $\mathrm{d}^2 R/\mathrm{d}t^2$. The latter can be found either by direct observation, or by measurement of ρ (and p) so that $\mathrm{d}^2 R/\mathrm{d}t^2$ can be calculated. As was mentioned in the last Unit, a direct measurement of $\mathrm{d}^2 R/\mathrm{d}t^2$ is difficult. So we should consider estimating the value of ρ.

The degree by which the actual age of a matter-dominated universe differs from the upper limit obtained from $1/H_0$ depends on the density as shown in Figure 7. The density is shown plotted as ρ_0/ρ_0^c, the ratio of the actual density to the calculated critical density.

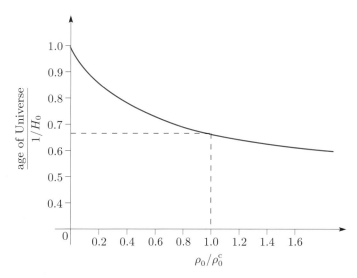

Figure 7   The ratio of the true age of a matter-dominated universe to the upper limit to its age (the latter being derived from the inverse of the Hubble parameter) for different assumed values of the present energy density, ρ_0.

$\left(\frac{dR}{dt}\right)^2 + kc^2 = \frac{8\pi G}{3c^2}\rho R^2$

Eq 5, p 10

As regards the shape of this curve, we note from Equation 5 that if $\rho = 0$, then dR/dt is a constant, so the graph of R in Figure 6 will be coincident with the dashed tangent line and thus the true age of the Universe will be equal to the upper limit given by the inverse of the Hubble parameter. Hence the curve of Figure 7 begins at a value for the ratio of 1.0. In SAQ 12 below, you will yourself be considering the especially interesting case where the density is critical. There you will be able to show that for $\rho_0/\rho_0^c = 1$, the ratio of the true age to the upper age limit is 2/3 as displayed in Figure 7. In general terms we see that the greater the density, the greater the deviation between the true age and its upper limit.

Study comment

The next two SAQs and their solutions should on no account be omitted.

Objectives 1, 3 & 4   **SAQ 12**   (a) Use Equation 5 to show that a matter-dominated, spatially flat universe must have a scale factor R satisfying

$$R\left(\frac{dR}{dt}\right)^2 = \text{ a constant.}$$

(b) Verify that

$$R = At^{2/3} \tag{9}$$

is a solution to the differential equation in part (a).

(c) Hence show that

$$t_0 = \frac{2}{3H_0} \tag{10}$$

for such a universe.

(d) If one were to assume that our actual Universe was of this type, what would be the current estimate of its age?

Objective 4   **SAQ 13**   Use the definition of q (Equation 14 of Unit 14) and the result you have just derived in SAQ 12(b), to prove that $q = 1/2$ for a matter-dominated, spatially flat universe.

Einstein–de Sitter cosmology

SAQs 12 and 13 in effect give the Friedmann–Robertson–Walker cosmology for a matter-dominated universe with $k = 0$. This particular case is conventionally called the standard Einstein–de Sitter cosmology. For many years it was thought to closely correspond to the cosmology of our actual Universe, and is therefore of particular interest. The development of such a universe is illustrated in Figure 8.

SAQs 12 and 13 led to important results. In particular, part (a) of SAQ 12 shows that

$$\frac{dR}{dt} \propto R^{-1/2}.$$

This means that the slope of the curve of R versus t must tend to zero as R tends to infinity. (This is not immediately obvious from Figure 8 because the convergence of $\mathrm{d}R/\mathrm{d}t$ towards zero is very slow.)

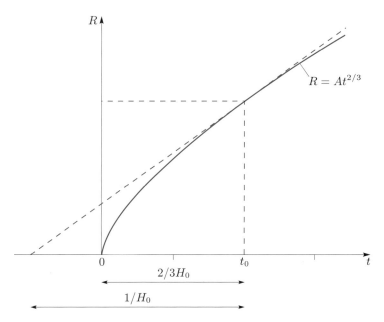

Figure 8   The variation of R with t for an Einstein–de Sitter universe. The slope approaches zero as R tends to infinity. The age of the universe is $2/3H_0$.

Part (c) of SAQ 12 underlined the fact that the actual age of the Universe will be less than the upper limit we have so far been dealing with — that obtained from $1/H_0$. It provides one of the plotted points on Figure 7 — that corresponding to $\rho_0/\rho_0^c = 1$. On the basis of Equation 10 so derived, we saw from part (d) of SAQ 12 that the current estimate for the age of a spatially flat, matter-dominated universe is 0.95×10^{10} years. Comparing this with what is currently thought by many astrophysicists as the lower limit to the age of the oldest stars, 1.2×10^{10} years, we see that there is a problem. We cannot have the stars older than the Universe itself! Clearly, something is wrong.

In searching for the source of the mismatch between Einstein–de Sitter cosmology ($\rho = \rho^c, \rho_\Lambda = 0, k = 0$) and astrophysical indicators of cosmic age, we certainly have plenty of candidates. The astrophysical methods used to establish the lower limit on the age are certainly open to question, though their broad agreement with one another looks difficult to challenge. Similarly one can question the current determinations of H_0, though these are being presented with increasing confidence by the astronomers responsible. There is also the specific assumption that $k = 0$, though you will learn of good supporting evidence for that (along with $\rho = \rho^c$) in the next section. That leaves us with the assumption $\rho_\Lambda = 0$, that we introduced at the beginning of Section 4.1. You saw in earlier units that both Type Ia supernova observations and attempts to explain the observed anisotropies in the cosmic microwave background radiation, indicated a non-zero value for ρ_Λ. So it is only natural to ask if that same assumption can lead to a revised 'best' model of the Universe that has an age more in keeping with the astrophysical limits.

The effect of a non-zero value for ρ_Λ was shown in Figure 3(b). It causes the graph of R against t to curve upwards during the era of dark-energy-dominance. If we were to go through the procedure of drawing a tangent to this upward-curving part of the graph, just as we drew a tangent to the downward-curving graph in Figure 8, then we would find the actual age of the Universe (t_0) was closer to the time represented by $1/H_0$ than in the Einstein–de Sitter case. This is just what is required to have the possibility of agreement between the cosmology and the astrophysics. However, in drawing the graphs it is important to realize that the observational parameter H_0 (which determines the gradient of the graph at $t = t_0$) must be the same in the two cases being compared. The effect of this is shown in Figure 9.

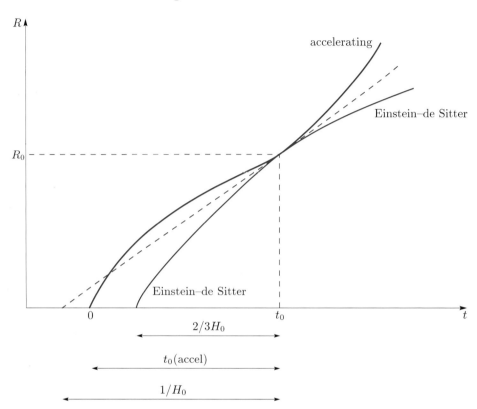

Figure 9   Schematic graphs of R against t for two $k = 0$ universes, sharing the same values of H_0, one (Einstein–de Sitter) with $\rho_\Lambda = 0$, the other (accelerating) with a positive value for ρ_Λ. The value of $1/H_0$ is the same in both cases, but in the case of the accelerating universe, it is closer to the time that the universe has been expanding (t_0).

Before leaving the age of the Universe there is one other body of observational results that should be noted. This relates to estimates of the age of the Universe based on efforts to model the angular power spectrum of the cosmic microwave background radiation (Unit 13, Section 6.3). This technique determines a wide range of cosmological parameters simultaneously. The values it delivers are not independently adjustable. Nonetheless, taking the comprehensive results from the WMAP satellite as a source, it is pleasing to note that this technique indicates an expansion age of $(1.37 \pm 0.02) \times 10^{10}$ years, as well as a positive value for ρ_Λ.

Objective 4   **SAQ 14**   Given that the cosmic microwave background radiation has been redshifted by $z = 1000$, and that the Universe is about 10^{10} years old, calculate the decoupling time (refer to Unit 13 Section 5.2) on the basis of the Einstein–de Sitter model of the Universe.

5 The density and composition of the Universe

The energy density of the Universe and the cosmic constituents that account for it are obviously of the utmost importance in cosmology. In discussing the age of the Universe it became very clear that we needed to know how closely ρ approached the critical value ρ^c that ensures a flat spatial geometry. Earlier, in Unit 13, you saw the importance of ρ_b, the density associated with baryonic matter (ordinary matter based on protons and neutrons), in determining the relative abundances of light nuclei. And this is to say nothing of the crucially important issue of dark energy and the value of ρ_Λ.

In this section, we shall discuss several aspects of the cosmic energy density, including the relative sizes of the contributions from dark energy (ρ_Λ), from matter of all kinds (ρ_{matter}), and from radiation $(\rho_{\text{radiation}})$. We shall also discuss specific contributions such as that due to baryons (ρ_b) and that due to the cosmic microwave background radiation. In keeping with the conventions of recent cosmological discussions, we shall make frequent use of a set of numerical quantities known as *density parameters*.

density parameters

These are, in principle, functions of time (though we shall often omit any reference to time to simplify notation) and are always defined (at any given time) by the ratio of the density of interest to the critical density at that time. Thus for example, the total density parameter of the Universe at time t is given by

$$\Omega(t) = \frac{\rho(t)}{\rho^c(t)} \qquad \text{where} \qquad \rho^c(t) = \frac{3H^2(t)c^2}{8\pi G} \quad \text{Eg 7 p16}$$

Similarly, for matter and for radiation

$$\Omega_{\text{matter}}(t) = \frac{\rho_{\text{matter}}(t)}{\rho^c(t)} \qquad \text{and} \qquad \Omega_{\text{radiation}}(t) = \frac{\rho_{\text{radiation}}(t)}{\rho^c(t)}$$

and for dark energy

$$\Omega_\Lambda(t) = \frac{\rho_\Lambda}{\rho^c(t)}$$

where ρ_Λ must be constant if it is to mimic the effect of Einstein's cosmological constant Λ. As usual we shall denote the present value of any of these density parameters by attaching a subscript 0, so for example the current value of the density parameter for dark energy will be denoted

$$\Omega_{\Lambda,0} = \Omega_\Lambda(t_0)$$
$$= \frac{\rho_\Lambda}{\rho^c(t_0)}.$$

On this occasion it makes sense to start our discussion with the values of the various density parameters obtained by the Wilkinson Microwave Anisotropy Probe (WMAP). We can then consider their implications in detail while comparing their values with other sources of information.

The WMAP density results, based on the simultaneous adjustment of several different cosmological parameters to achieve the best possible agreement between a theoretically determined angular power spectrum and that determined from direct observations are as follows:

Total density parameter: $\Omega_0 = 1.02 \pm 0.02$

Dark-energy density parameter: $\Omega_{\Lambda,0} = 0.73 \pm 0.04$

Matter density parameter: $\Omega_{\text{matter},0} = 0.27 \pm 0.04$

Baryonic density parameter: $\Omega_{\text{b},0} = 0.044 \pm 0.004.$

The density parameter for radiation is not included amongst these results, but is believed to be very small at present and is mainly attributable to the cosmic microwave background. A reasonable estimate is

Radiation density parameter: $\Omega_{\text{radiation},0} = 0.000\,05.$

As you can see, these results indicate that dark energy is not only present, but that it accounts for slightly more than 70% of the energy density of the Universe. They also indicate that almost all of the remainder is due to matter, though baryonic matter (atoms and molecules etc.) only accounts for a small part (about one-sixth) of that matter. Very significantly, the WMAP result for the total density parameter is consistent with 1, in keeping with a number of results from other sources, which gives good support to a widely held theoretical prejudice in favour of a $k = 0$ Universe with a flat spatial geometry.

Now let us turn to the astronomical evidence regarding density that will occupy us for much of this section. The situation regarding dark energy has already been substantially discussed in Unit 13, where we considered the impact of the results from Type Ia supernova measurements. For that reason, we shall turn immediately to the distribution of matter and its associated mass, even though matter is thought to be only the minor contributor to the average energy density.

We shall find in what follows that if we add up the masses of all the luminous components of galaxies — mainly the stars — out to any large distance (as determined by the methods outlined in Unit 13), and if we divide by the volume out to that distance, then we get a mass density that is something like $\rho_{\text{matter},0}/c^2 \approx 2 \times 10^{-29}\ \text{kg m}^{-3}$. This is nearly three orders of magnitude below the critical mass density, $\rho_0^{\text{c}}/c^2 \approx 1 \times 10^{-26}\ \text{kg m}^{-3}$. However, we shall also find that the stars account for only a small fraction of the matter in the Universe. There is evidence for at least 300 times more mass within, around, and between galaxies than in the constituents we can see; most of the Universe is 'dark'.

dark matter

The composition of this *dark matter* is largely a mystery. The identification of its exact nature is one of the most pressing subjects of current research. It is not thought that there is any direct link between dark matter and dark energy. The term 'dark' is used in both cases simply to indicate that its existence is deduced indirectly, through its influence on entities that we can directly observe.

5.1 The distribution of mass

We begin our survey of the density of the Universe by counting up the mass for which we have direct evidence. It is natural to start by thinking that most of the mass is to be found in the stars. The Sun is a star, so what is its mass?

(a) *The mass of the Sun*

We measure the mass of the Sun, and of the Earth and other planets, by *weighing* them. When we weigh ourselves on the bathroom scales, we use

★ p24
|g| = GMₒ/Rₒ²

The subscript ⊕ is the conventional notation for quantities that refer to the Earth.

the gravitational force of the Earth on our bodies to infer our mass: $\mathbf{F} = m\mathbf{g}$, so $m = |\mathbf{F}|/|\mathbf{g}|$. Astronomers use a related technique to measure the mass of the Earth itself. The magnitude of the acceleration due to gravity, g, is given by $GM_\oplus/R_\oplus^2$, (see Unit 2).★ If we measure g for a freely falling object in the laboratory, and if we know the radius $R_\oplus$ of the Earth, then we find the mass of the Earth,

$$M_\oplus = \frac{gR_\oplus^2}{G}.$$

In practice, the best way to measure the acceleration due to gravity is to observe the acceleration of a satellite orbiting the Earth, because it stays up for so long that we can get a very accurate measurement of its motion and orbit. As we saw in Unit 2, for a circular orbit of radius r (measured from the centre of the gravitating body), the orbital speed v is related to the satellite's acceleration a by $v^2 = ra$. The mass of the gravitating body is given by a formula similar to the previous equation:

$$M = \frac{ar^2}{G} = \frac{v^2 r}{G}. \tag{11}$$

We can use this same formula to determine the Sun's mass from the orbit of one of its satellites, such as the Earth itself. The radius of the Earth's orbit is 1 AU (astronomical unit), or 1.496×10^{11} m. Knowing that the Earth takes 1 year to circle the Sun, it follows that the mass of the Sun is given by

$$M_\odot = 1.989 \times 10^{30} \text{ kg}.$$

Objective 5   **SAQ 15**   With the data provided, verify the above result.

It is interesting to note that astronomers can measure the radius and period of the Earth's orbit to very high accuracy, exceeding 9 decimal places. But experiments can measure G accurately to only 4 or 5 decimal places, so this uncertainty sets the limit on the accuracy with which we know the mass of the Sun.

(b) *The mass of the Galaxy*

If we assume that the Galaxy is made up mostly of stars, then we can estimate its mass by counting the stars in it. There are too many to count individually, but we can estimate their number by measuring the total light they emit. The total luminosity of the Galaxy is about 4×10^{36} W. This is about 10^{10} times the luminosity of the Sun — the Sun being a fairly average star. Thus, we might conclude that the Galaxy contains about 10^{10} stars and has a mass of about $10^{10} M_\odot$.

★ Solar system ? yes

dynamical method for mass measurement

We can check this estimate by using Equation 11. The stars near us in the Galaxy seem to be moving on a circular orbit around the centre of the Galaxy. Our★ speed is about 220 km s^{-1} and the orbital radius is about 2.5×10^{20} m. It follows from Equation 11, that the mass of the part of the Galaxy contained within the Sun's orbit is about 2×10^{41} kg, or $10^{11} M_\odot$. This method, which is called the *dynamical method* for measuring masses, measures *all* the mass, not just that in luminous stars. We conclude, therefore, that there is at least ten times as much matter in the Galaxy as there is in its individual stars.

mass-to-light ratio

A convenient way of speaking about this is to say that the Galaxy has a ratio of mass to luminosity that is about 10 times that of the Sun itself: its mass is 10^{11} solar masses but its luminosity is only 10^{10} solar luminosities. Astronomers have a compact way of expressing this: they say that the *mass-to-light ratio* of the Galaxy is about 10. Mass-to-light ratios are always quoted on a scale on which the Sun is 1.

So far, our argument has told us only about the mass of the Galaxy contained *inside* the orbit of the Sun. The dynamical method does not measure mass outside our orbit, since any uniformly distributed mass outside has no gravitational effect on us. However, we are fairly far out in the Galaxy, and most of the luminous stars are inside our orbit. (We might also mention that Equation 11 assumes, strictly speaking, that the mass of the object being weighed is distributed spherically about its centre. This is certainly not true of our Galaxy, which has spiral arms like our nearest neighbour Andromeda. But the non-spherical distribution of mass probably does not make more than a 50% error, and this is good enough for us: we are dealing with a factor of 10 in missing mass.)

(c) *Luminous regions of other galaxies: estimating the visible mass density of the Universe*

If we assume that our Galaxy is typical of all galaxies, so that other galaxies have a mixture of luminous stars and dark matter that is similar to ours, then we can estimate the masses of other galaxies by measuring their luminosities and multiplying by the same mass-to-light ratio of 10. In this way we can infer that the Andromeda galaxy (see Figure 10) has a mass very similar to that of our own Galaxy, i.e. it lies in the range 1–$2 \times 10^{11} M_{\odot}$.

As with our Galaxy, we can try to test this with the dynamical method. To do this we have to find something that goes in orbit around the galaxy whose mass we want to measure, and we have to be able to measure the orbital speed. Spiral galaxies (like our own and Andromeda) are good candidates for this method, because they consist of stars that go in orbits around the galactic centres. While we cannot measure individual stars at such great distances, we can look at the light emitted by all the stars in a given region of the galaxy's disc, and determine whether it exhibits a redshift/blueshift indicating that the stars in that region share a systematic orbital velocity about the galactic centre. We can then examine how such a velocity changes with distance from the centre of the galaxy. The shift of spectral lines measures the speed of the stars along the line of sight, i.e. directly toward us (blueshift) or away from us (redshift). We have no way of estimating the speeds of objects *across* the line of sight. For example, if a spiral galaxy is oriented perfectly face-on toward us, so that its outline is a circle, then its rotational speed is across the line of sight, and it does not affect the redshift we measure anywhere in its image (assuming relativistic time dilation effects are negligible).

We could in principle wait for the spiral pattern of a galaxy to rotate around the line of sight, but this would typically take 10^8 years, too long to wait!

But if the galaxy is inclined to the line of sight, so that its shape seems elliptical, as in Figure 10, then a component of the rotational velocity of the stars will be along the line of sight, and will contribute a redshift or blueshift. We would expect in such a case that the stars on one side of the image of the galaxy should be moving away from us relative to the centre of the galaxy, and those on the other side should be moving toward us in their orbits. The redshift should be greater on one side of the image than on the other.

If we assume that the galaxy is circular, then we can use its *apparent* shape to estimate how much of the rotational velocity is projected along the line of sight, and thereby infer from the differential blueshift what the true rotational speed of the galaxy is. This is illustrated in Figure 10, which shows orbital velocity measurements of the Andromeda galaxy.

When astronomers survey galaxy rotation in the visible parts of galaxies, such as the inner region in Figure 10, they find rotational speeds and radial distances that are consistent with what we see in our Galaxy. This confirms that the visible regions of galaxies like ours typically have mass-to-light ratios of about 10. Unfortunately, the dynamical method only works on galaxies that are near enough to ours to allow astronomers to measure the changes in spectral shifts across the image; most distant galaxies have too small an image for this. But since the dynamical method agrees with the mass-to-light method for the visible parts of nearby galaxies, astronomers generally feel confident in using the mass-to-light method to estimate the masses of the luminous parts of all other galaxies.

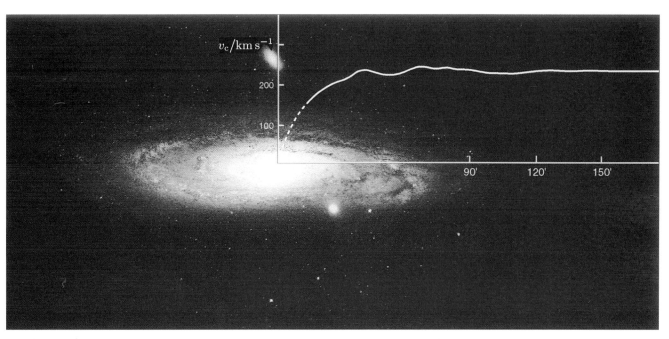

Figure 10   Rotational velocities of the Andromeda galaxy superimposed on a photograph of the same galaxy on the same scale. The horizontal scale on the graph gives the apparent angular size in arc minutes. The velocity measurements are based on redshifts in the radio spectrum of the galaxy.

By determining the distances of galaxies from their overall redshifts, and by estimating their masses from their luminosities using the mass-to-light method, astronomers arrive at a mass density that is about 2% of the critical mass density. Astronomers call this the *visible mass density*. By this they mean the mass contained in the visible regions of galaxies. As we have seen, 10% of this mass (i.e. 0.2% of critical density) comes from matter that is actually radiating light (stars), while 90% of it belongs to a form of dark matter. This dark matter seems to be intimately associated with the luminous stars, inasmuch as it is found in the same proportion to stars in the visible portions of most spiral galaxies.

Objective 5   **SAQ 16**   The critical density is $\rho_0^c/c^2 \approx 1 \times 10^{-26}$ kg m^{-3}, and we have just seen that the density of the visible parts of galaxies, when averaged over large volumes of the Universe, is only 2% of this. If we take the average mass of the visible part of a galaxy to be $10^{11} M_\odot = 2 \times 10^{41}$ kg, then what is the average density (number per unit volume) of visible galaxies themselves? How large is the volume of space that contains a single galaxy, on average? If this volume is roughly shaped like a sphere, what is its radius? From this, deduce the average distance between visible galaxies. How many times larger is this than the typical size of the visible part of a galaxy, which is 3×10^{20} m? The distance from our Galaxy to the Andromeda galaxy is about 1.5×10^{22} m. Is this an average separation, or are we exceptionally near to our neighbour?

(d) *Galactic haloes*

If it was salutary to learn that 90% of the matter contained in the visible disc of a typical galaxy is unseen, a further surprise comes when we apply the dynamical method to estimate the mass of a spiral galaxy *outside* the visible disc. There are of course no visible stars that we can measure there, but it turns out that many spiral galaxies still have a considerable amount of hydrogen gas in orbit out there, where none of it has been turned into visible stars. Radio telescopes can detect this gas and measure some of its spectral lines. Figure 10 shows the rotational velocity of the Andromeda galaxy as measured by radio observations. The velocity appears to level off outside the visible image and continues at a constant value to a very large radius.

This implies that there must be a substantial amount of mass outside the visible region. If all the mass were contained in the visible disc, then the orbital velocity would decrease with distance, according to Kepler's laws. Instead, we can see from Equation 11 that if v remains constant as the radius increases, then M increases *linearly* with radius. Indeed, the radio observations of Andromeda extend out to at least 2.5 times the visible radius, and this implies that the overall mass of Andromeda is at least 2.5 times the mass we infer from looking at its visible region. Astronomers call this extra mass the *halo* of the galaxy. Practically every spiral galaxy that is near enough for radio astronomers to measure the hydrogen orbital velocity shows a large halo, contributing 2 to 5 times as much mass as the visible part of the galaxy.

The hydrogen gas itself does not contribute much to this mass. The radio observations tell us that there is very little of it: it is merely a tracer of the distribution of some other, unknown constituent of the Universe. The hydrogen tells us that the mass continues to increase in proportion to radius as far out as we can measure the hydrogen. What happens beyond that?

To see how far haloes extend, we can consider *pairs* of galaxies that are bound to one another, and try to use Equation 11 for their orbital motion. This is another version of the dynamical method, but here we take the galaxies themselves to be the orbiting bodies. Unfortunately, the method becomes difficult to use at this point. The evidence is less certain since it is not likely that pairs of galaxies will orbit each other on perfectly circular orbits — which is one of the assumptions of Equation 11. But, by making such measurements on many pairs of galaxies, one can build up a statistical picture of their haloes, averaging over the uncertainties of each individual case. The evidence is very suggestive that galaxy haloes typically extend to at least 10 times the visible disc, containing at least 10 times the mass of the visible part. This means that *the mass-to-light ratio of galaxies goes up to 100 when their haloes are included.* This pushes the mass density of the Universe up to about 20% of the critical density needed to close the Universe.

(e) *Dark matter between galaxies*

Can we find evidence for yet more mass — additional mass that would take us even closer to the critical value for the density?

We have seen that galaxies usually clump into groups, called clusters. Although the distances between galaxies can be vast, there seems no doubt that these groups are held together by their mutual gravitational attraction, so that it is in principle possible to use the dynamical method to estimate the total mass of a cluster. The difficulty is that the individual orbits of galaxies are not known, and so it is not possible to use Equation 11 directly. Instead, we look for clusters containing so many galaxies that it is possible to average over their individual motions and again estimate the mass in a statistical way.

Galaxies often group into rich clusters, containing many thousands of individual galaxies. In such swarms, the galaxies will have different random velocities, much as the atoms in a gas do. But taken on average, it can be shown that the spread of random velocities (called the *velocity dispersion*) is a good measure of the self-gravity of the cluster; the faster the random velocities, the more mass there must be to hold the cluster together.

Figure 11   The Coma cluster.

When this is done for a nearby rich cluster, the Coma cluster (see Figure 11), astronomers find that the central regions have a mass-to-light ratio of about 240, instead of the previous value of 100 we obtained for a single galaxy (including its halo). What is more, the measurements have been done for many clusters, always with the similar result: the mass-to-light ratio of rich clusters is systematically larger than that for individual galaxies by a factor of between 2 and 4. Clusters evidently contain extra dark matter that is not held in the haloes of their galaxies. It lies in the vast spaces between galaxies. We shall see in Section 5.2 that astronomers now have direct observations of X-rays emitted by some of this material, so in this sense not all of it is completely 'dark'.

There is another method that gives independent evidence of cluster masses. This one is based on *gravitational lensing* — introduced in Unit 12. When the line of sight to a distant galaxy or quasar passes through a rich cluster of galaxies, the gravitational field of the cluster can produce multiple images or distorted images of the distant object. By carefully measuring the distortions and modelling the distribution of mass in the lensing cluster, astronomers can make estimates of the total mass in the cluster that acts as the lens. The resulting mass estimates are consistent with those from galaxy velocities in clusters. All clusters that have been examined in this way, provided they are larger than about 3×10^{22} m across, show a similar mass-to-light ratio of about 240.

With this evidence, then, it is tempting to say that the mean density of the Universe is at least 240 times the value we infer for the visible components of galaxies. This would imply a matter density of at least $\rho_{\text{matter},0}/c^2 \approx 5 \times 10^{-27} \text{kg m}^{-3}$, 50% of the critical density.

However, things are not quite so simple. A major problem in the use of mass-to-light ratios arises from observational bias that is hard to remove. In particular, it has been suggested that the clusters of galaxies used to estimate mass-to-light ratio on large scales may give an unrepresentative value because their density causes them to contain more highly evolved stars than would be true on average. Attempting to allow for this leads to an estimate of $\rho_{\text{matter},0}$ that is closer to $2 \times 10^{-27} \text{kg m}^{-3}$, though the uncertainties are probably quite substantial. The overall conclusion seems to be that mass-to-light ratios give results that are within the limits indicated by WMAP , though there is plenty of room for further refinement and hence the potential for future conflict. What is certain is that the growth of mass-to-light ratios with increasing size scales provides yet more evidence that much of the matter is dark matter.

5.2   Some types of matter and radiation

We have seen that, whatever its true value, the mass-to-light ratio of the Universe is much greater than 1, so stars make up only a small fraction (probably less than 1%) of the total matter in the Universe. What is the rest made of? In this section we shall go through a list of possible types of matter and radiation, and we shall discover what bounds astronomers are able to place on them. We shall find, remarkably perhaps, that ordinary matter (protons, neutrons, and electrons) cannot account for all the matter that we measure in clusters. The hunt for the remaining dark matter is one of the major enterprises in astronomy today.

5.2.1   Baryonic dark matter

As mentioned earlier, protons and neutrons are *baryons*, and they account for almost all the mass of ordinary stars. (Electrons, the remaining constituent of stars, are so light that they make up less than 0.1% of the mass of a star.) It would be natural to ask if there is 'dark baryonic matter' as well — baryonic matter that is not radiating energy in the way that it does in stars. There are many ways that baryons could remain dark, and we shall go through some of them now, describing the evidence for them. Then we shall turn to a more fundamental issue, namely, that both the WMAP results and predictions of primordial abundances of light nuclei described in Unit 13 show that *the Universe cannot contain enough baryons to account for all the dark matter.*

(a) *Cool matter: gas, dust, and remnants*

Baryons have to be hot to radiate visible light. The first place to look for dark baryons, therefore, is in cooler places. The visible part of our Galaxy hides 10 times as much mass as we can see in luminous stars. Could this be in cool baryonic matter?

Figure 12   Clouds of interstellar gas and dust photographed by the Hubble Space Telescope.

This ratio of 10 is much larger than we can easily account for. Early generations of stars may have burnt out and left behind black holes or neutron stars, or white dwarfs that have cooled off to invisibility. But since the age of our Sun, which is a typical star, is almost half the age of the Galaxy, probably not much more than half of the mass of the Galaxy can be in these remnants of stellar evolution. There are also vast clouds of interstellar gas and dust in the Galaxy, which are the places where new stars are forming (see Figure 12). Some of these are visible where the dust in them obscures more distant stars or where newly formed stars shine brightly. Other, cooler, more diffuse clouds of hydrogen radiate at a wavelength of 21 cm, which astronomers detect with radio telescopes. But all this gas probably doesn't account for more than 10% of the mass of ordinary stars. So even within galaxies astronomers cannot account in obvious ways for the dark matter.

(b) *MACHOs*

As we have noted, the haloes of galaxies hide even more dark matter, perhaps 5–10 times as much as in the visible portions. If that is in the form of baryons, then one possibility is that it is gathered together as compact objects such as small stars or black holes, left over from the first stages of galaxy formation. Astronomers have come to call these MACHOs, which stands for MAssive Compact Halo Objects.

Halo objects must be of low or essentially zero luminosity, or we would see them in telescopes. There are two kinds of object that might qualify: brown dwarfs and black holes (the latter formed from what was originally baryonic matter). Brown dwarfs are objects that have such small mass that they have never reached temperatures high enough to ignite nuclear reactions. Their mass must be less than about $0.08 M_\odot$. However, if there were enough of them they could make up the dark matter. Black holes in the halo presumably must be much more massive than the Sun, perhaps $10 M_\odot$ or more. They could in principle have any mass, even up to $1000 M_\odot$. Again, the halo could be made of these without their having any noticeable effect on ordinary astronomical observations.

Since black holes 'have no hair' one cannot attribute a particular number of baryons to them; our classification of black holes under 'baryonic matter' is somewhat arbitrary, but possibly justified by their history!

How could such objects be distributed in such a different way from ordinary stars? The answer is not known, but presumably it would reflect the way the Galaxy took shape in the early stages. Perhaps as the original cloud of gas that formed the Galaxy began to collapse inwards, most of it became trapped in a first generation of stars, either unusually small stars (brown dwarfs) or unusually massive ones (which evolved into black holes). Because they formed before the collapse was finished, they were left distributed over a much larger volume of space. They now move randomly through this volume, so some of them pass through the disc. But if they are hard to see then we would not have noticed them, even nearby.

Astronomers have now detected some of these objects, most likely brown dwarfs. The method of detection is remarkable: it uses gravitational lensing. If a MACHO happens by chance to pass across our line of sight to a more distant star, then it can do two things to the image of the star: it can displace it (by bending the path of the light) and it can magnify its intensity (by focusing the light). The displacements expected in this case are too small to measure, but astronomers can look for a sudden apparent brightening of a star.

The unique signature of such a *microlensing* event is that the brightening should be (i) symmetrical in time (the star should get dimmer in exactly the reverse way to that by which it brightened), and (ii) independent of wavelength in the star's spectrum. These features distinguish microlensed events from ordinary stellar variability. When a star brightens up because of some internal mechanism, it rarely affects all wavelengths in the same way. Also, the subsequent dimming is usually very different from (and slower than) the brightening. In addition, microlensing events should be isolated: a given star is unlikely to show microlensing twice, while a variable star might well brighten repeatedly. Finally, we note that the time-scale of the brightening tells us whether we are seeing a brown dwarf or a black hole.

To have a chance to see such events, one must monitor millions of background stars, looking for the rare one that happens to be lensed. With modern computer-controlled telescopes and CCD detectors, two groups of astronomers have made such observations. They have found a few clear examples of microlensing by brown dwarfs in the direction of the Large Magellanic Cloud (LMC). Light from the stars of the LMC passes through our galactic halo, and occasionally passes near enough to a brown dwarf in the halo for the image of the LMC star to be briefly intensified (see Figure 13). From the number of observed events, it appears that brown dwarfs in the halo constitute a total mass that is less than a few times $10^{11} M_{\odot}$. Thus, brown dwarfs may be an important component of the dark matter that surrounds galaxies like our own, but they cannot be the only component.

CCD: charge-coupled device. CCDs are solid-state detectors, replacing photographic film for many astronomical purposes at visible and near-visible frequencies.

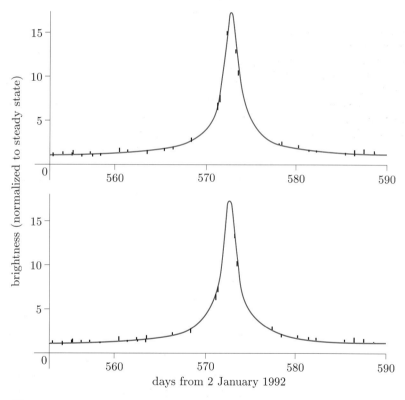

Figure 13 A microlensing event: the peaks in brightness are produced by a passing MACHO. The effect has been observed independently in red and blue light, hence the two graphs.

(c) *Hot gas*

Another possibility is that baryonic dark matter is distributed as a hot, thin gas. If it were not hot, it would clump and we would see the resulting clouds. If it is sufficiently hot, it is not bound to individual galaxies, but instead is part of the general distribution of dark matter in clusters.

Evidence for this dark matter comes from X-ray observations. Satellites containing X-ray telescopes have found intense X-ray emission from clusters. In Figure 14 we show contour lines of the emission from the

Coma cluster. It comes from all parts of the cluster, and is not associated with any particular galaxy. Astronomers conclude that this cluster must contain gas that is very hot, about 10^8 K, in order to produce this radiation. What heats this gas is not at all clear, but the gas itself must be ionized hydrogen, and it indicates a major repository for baryons.

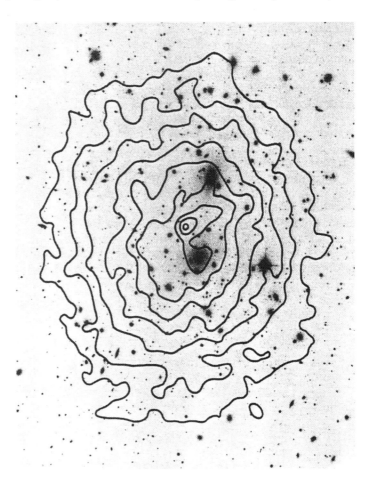

Figure 14 Contour lines showing the intensity of X-ray emission from the Coma cluster.

However, there is not enough of it to account for all the dark matter in the Coma cluster. From the intensity of the X-ray emission, it is estimated that the hot gas makes up only about 10 to 20% of the total mass of the cluster. So at least 80% of the cluster mass is still unidentified.

(d) *Limits on further forms of baryonic matter in the Universe*

We have in mind black holes formed by collapsing baryonic systems. Black holes cannot themselves be described as baryonic due to their 'hairlessness'.

Could baryonic matter, in some other form not yet identified, such as very massive black holes, make up the remaining dark matter? One argument that this is unlikely to be the case comes from the primordial abundances of the elements emerging from the big bang.

You will recall from Unit 13 that the primordial abundances were sensitive to the present value of the baryonic density and indicated a value for ρ_b/c^2 of a few times 10^{-28} kg m^{-3}. This is very much in keeping with the WMAP results which imply $\rho_b/c^2 \approx 4 \times 10^{-28}$ kg m^{-3}. Together, these two results put stringent limits on further forms of baryonic matter.

Objective 5 **SAQ 17** (a) Find k/R^2, taking the value of the present mass density to be $1.0 \times 10^{-27}\,\mathrm{kg\,m^{-3}}$, and assuming H_0 to be $2.3 \times 10^{-18}\,\mathrm{s^{-1}}$.

(b) What value of the total mass density would make $k = 0$?

Objective 5 **SAQ 18** Using the values given in Unit 13, Section 7.2, compare the estimated deuterium (^{2}H) mass fraction with that which would be predicted if baryons accounted for all the density in a universe with flat spatial geometry.

Objective 5 **SAQ 19** Helium is abundant and has survived from the big bang. Why is it not used to estimate the density of the Universe?

5.2.2 Electromagnetic radiation

What about other, non-baryonic, contributions to the energy density of the Universe?

The very starlight by which galaxies are observed carries energy that contributes to the self-gravitation of the Universe. This applies, of course, to radiation throughout the electromagnetic spectrum. Could it account for a major part of the density of the Universe? There are significant amounts of radiation in many wavebands:

(i) 21 cm radio waves from cool neutral hydrogen;

(ii) continuous-spectrum radio waves generated by charged particles in magnetic fields;

(iii) the cosmic microwave background radiation generated in the early Universe;

(iv) infrared, optical, and ultraviolet light from stars;

(v) X-rays from gas heated by falling onto very dense stars and from clusters of galaxies;

(vi) gamma-rays from nuclear interactions.

Of all these kinds of radiation, the one that has the largest energy density and dominates all other contributions is the microwave background. Its energy density is that of a 2.7 K black body spectrum, which is about $10^{-14}\,\mathrm{J\,m^{-3}}$. This is only about 10^{-5} of the critical density of the Universe. This simple number has several important lessons to teach us.

The first lesson is that the energy density of the visible matter in the Universe is about 1000 times larger than that of the cosmic microwave background today. This was not always the case. As we saw in Equations 3 and 4, the energy density due to matter varies over time as $1/R^3$ whereas that due to radiation varies as $1/R^4$. Up to 10 000 years after the big bang, it was radiation that contributed the greater energy density.

The second lesson is that there are many more photons than baryons in the Universe. At a temperature of $T = 2.7$ K, the typical photon energy is about 4×10^{-23} J (or 2×10^{-4} eV). The velocity of the typical baryon being much less than c, its energy is approximately mc^2, namely about 1.5×10^{-10} J, and this is a factor of 4×10^{12} larger. Since the total energy density associated with baryons is only a factor of 1000 larger than that of

photon-to-baryon ratio

the photons, *there must be many more photons per unit volume than baryons, by a factor of* 4×10^9. This *photon-to-baryon ratio* has probably not changed very much since the stage of baryon formation in the very early Universe. Then there was an almost perfectly balanced mixture of matter and antimatter. Most of this annihilated, producing lots of photons, and leaving a small excess of baryons over antibaryons. The ratio tells us that there was an imbalance between matter and antimatter at very early times of a few parts in 10^9. This is one of the fundamental facts about the early Universe that can guide physicists who are trying to construct theories of physics at high energies: there must be something in the laws of physics that slightly prefers producing matter instead of antimatter in nuclear reactions, but only at the level of 1 part in 10^9.

The third lesson is that we should not be surprised that the microwave background contains more energy than starlight or other forms of radiation. Most radiation is produced by nuclear reactions involving baryons, and these typically cannot convert more than 0.1% of the mass of the baryons into energy that gets radiated away. Even if every baryon in the Universe went through such reactions (which has not happened, since most of the visible Universe is still pure hydrogen), they would only have produced radiation with at most 0.1% of the energy density of the baryons, which is the same as that of the cosmic microwave background radiation.

The fourth lesson is the simplest: electromagnetic energy cannot boost us to the critical density today.

Objective 5   **SAQ 20**   Calculate the fraction of mass that is converted into other forms of energy in the reaction leading to deuterium

$$p + n \rightarrow {}^2H$$

where the mass of the proton is 1.6725×10^{-27} kg, that of the neutron is 1.6748×10^{-27} kg, and that of the deuteron is 3.3433×10^{-27} kg.

5.2.3   Gravitational radiation

As we saw in Unit 12, gravitational waves are emitted by all accelerating masses (apart from the special case of matter accelerating in a spherically symmetric fashion). However, the waves are usually very weak. Strong backgrounds of gravitational waves, however, could have been produced in the early Universe. If gravitational radiation was ever in equilibrium with electromagnetic radiation immediately after the big bang, then its energy density today would be a little smaller than that of the microwave background, i.e. less than 10^{-5} of the critical density. Theories of inflation and of cosmic defects, which we shall discuss in Section 6, also predict a gravitational wave background, but one that must be well below the critical density.

There are observational limits on the background from studies of pulsars (mentioned in Unit 12). Some pulsars are so regular that, over periods of years, they keep time to accuracies as good as those of the best atomic clocks. If a gravitational wave passes between a pulsar and the Earth, the changing gravitational field will change the arrival times of the radio waves from the pulsar. By monitoring the small observed irregularities in pulsar timing, limits have been set that show that the gravitational wave background has an energy density less than 10^{-6} of the critical density, at least at very low gravitational wave frequencies. There is the possibility that the background could be larger at higher frequencies, but probably not much larger than 10^{-4} of the critical density, because otherwise it would disrupt the nucleosynthesis. Purpose-built gravitational wave detectors may soon be able to measure the background if it is larger than about 10^{-10} of the critical density.

5.2.4   Neutrinos

Many nuclear decays produce neutrinos. Unlike photons, which scatter from charged particles, neutrinos do not interact with electric charge. They are elusive and hard to detect. Because of this, their physics is still rather poorly understood. One uncertainty is their mass. Physicists at first thought that they were massless particles, travelling at the speed of light, the same as photons. It is now clear that they have mass, but its value has not yet been accurately determined.

Another uncertainty about neutrinos was how many kinds there are. Neutrinos are leptons, a class of particles that includes electrons. Physicists have known for some time that there are at least 3 families of leptons. The first family consists of the electron and the 'ordinary' neutrino or electron-neutrino, associated with it in beta decay. The second family comprises the muon (the heavy form of electron you met in Block 2) and an associated new kind of neutrino, the muon-neutrino. In 1975, an even heavier electron-type particle, the tauon (τ), was discovered, and it too has an associated 'tau-neutrino'. This leads to the obvious question: how many families of leptons are there?

The theory of big bang nucleosynthesis provided the first reliable answer to this: it predicted that there should be exactly 3 families of leptons. Particle physics experiments at CERN subsequently verified this. This was a significant triumph for the big bang model. Here is how the deduction was made:

The number of kinds of neutrinos affects the way nucleosynthesis proceeds in the expanding universe. Each kind of neutrino formed a separate 'gas' that shared the total energy in any particular region with the others. When the Universe expanded and cooled to the right temperature for elements to form, each kind of neutrino gas had to have the energy of a gas at that temperature, so if there were more kinds of neutrinos, then there was more energy in the neutrinos, and they exerted more self-gravitation to decelerate the expanding Universe. In order to reach the present density and expansion rate of the Universe, a model with more families of neutrinos had to be expanding faster at the time of nucleosynthesis. But this faster expansion would have quenched the nuclear reactions faster, leading to changes in the abundances of elements. The observed abundances of lithium and deuterium can only be fitted adequately if the Universe had only 3 kinds of neutrinos.

A few years after astronomers made this prediction about neutrino physics, particle physics experiments at CERN discovered a particle called the Z^0. The Z^0 particle can only decay into leptons, and the more families of leptons there are, the faster it can decay. The measured decay rate of the particle confirms that there are only 3 families of leptons.

This vindication of the big bang model gives us added confidence that the limits on baryon density we quoted in Section 5.2.1 are also correct: at about 5% of the critical density of the Universe.

But are the neutrinos *themselves* an important form of dark matter? The Universe must contain vast numbers of them. However, their contribution to the unseen mass of the Universe depends crucially on whether the neutrino has a significant mass. Earlier we mentioned that, until recently, they were assumed to be massless. But this was not correct.

In the very early Universe, when matter was so dense that even neutrinos scattered frequently, there were roughly equal numbers of neutrinos and photons: the neutrino gas and the photons had the same energy density and therefore the same number of particles. The typical energies at that time were far in excess of the few eV that mark the upper limit on the possible mass energy for the neutrinos, so neutrinos behaved like massless particles, i.e. like photons. In particular, their energy density decreased at the rate of $1/R^4$, in step with that of the photons as the Universe expanded and the overall temperature decreased. In the case of neutrinos in the present Universe, if the mc^2 contribution to their energy is still small compared to their kinetic energy, then each neutrino will typically have an energy similar to that of a typical photon in the present-day cosmic microwave background, i.e. about 2×10^{-4} eV, and the neutrino energy density will still be behaving as $1/R^4$ and comparable in magnitude with that of the photons. From this we can conclude that neutrinos will not make a significant contribution to the dark mass of the Universe.

But if their mass is much larger than 10^{-4} eV/c^2, say the 1 or 2 eV/c^2 that some physicists believe is reasonable, then at some time in the past the point was reached where mc^2 became the major contributor to their energy. After that point, just as was the case with baryons, their energy (now due mainly to their mass rather than their kinetic energy) did not decrease significantly as the Universe continued to expand. (The energy *density* of neutrinos and baryons does, of course, decrease as the Universe expands. This is because the volume containing any group of them increases. But the energy of each particle does not change much.)

Note that the number of neutrinos remains comparable to that of the photons, not to the baryons. This arises from the fact that, unlike baryons, neutrinos and antineutrinos do not interact strongly enough to have annihilated one another. Therefore there are more than 10^9 neutrinos for every baryon. Since baryons make up about 4 or 5% of the critical density, if the mass of the neutrino compared to that of the baryon had been as large as 2×10^{-8} (i.e. 20 eV/c^2), their energy density would have been comparable to the critical density.

Their contribution is almost certainly less than one-tenth of this, but they might still be significant.

5.2.5   Exotic particles

If neutrinos and baryons do not make up the dark matter, then it must be made of particles that physicists have not yet discovered in laboratory experiments. What kinds of particle could be candidates? The particles must have been comparatively inactive at the time of nucleosynthesis, or they would have disturbed the present helium and deuterium abundances. There are two ways that such particles could provide the critical density today:

(i) The particles have a rest energy above the mean energy which photons possessed at the time of nucleosynthesis (about 2 MeV), but were of low abundance. This being so, they might not have provided significant mass at the time helium formed, but could provide significant mass today — now that the photons have redshifted to low energies. Such exotic particles might be very massive fundamental particles yet to be discovered. Alternatively, they might even be the primordial black holes of Unit 12.

(ii) The particles have a low mass, but interact with other particles even more weakly than neutrinos do. Under these circumstances they could have 'dropped out' of the nuclear reactions in the Universe well before the epoch of nucleosynthesis. They would also have had to have a relatively low abundance in order not to change the overall expansion rate of the Universe at the time of nucleosynthesis, for this would have altered the observed abundances. The particles might be ones that have already been predicted by some model theories of high-energy physics. Astronomers call them WIMPs, which stands for Weakly Interacting Massive Particles. (Note that in this context the term 'massive' simply means that the particles have a mass; it does not imply that it is necessarily large.)

In both cases, the particles would now have low abundance compared to photons. To contribute significantly to the dark matter, their mass would have to be well above the value of $20\,\mathrm{eV}/c^2$ we deduced for the far more abundant neutrinos.

It is interesting that WIMPs are stronger than MACHOs as dark-matter candidates these days.

5.3   Hot and cold dark matter: helping galaxies to form?

We have discussed the evidence for dark matter in terms of its present gravitational effect on galaxies, and in terms of candidates for its constituent particles from theory and experiment in particle physics. There is however another kind of evidence: the gravitational effect of the dark matter was probably necessary in order to cause galaxies to form in the first place, and the masses of dark-matter particles would have had a significant effect on the way galaxies have clustered.

Stars and galaxies are such an obvious feature of the Universe that it comes as a surprise that we still do not understand how they could have formed. Galaxies must have condensed from the expanding Universe by their own self-gravity, and that means they must have formed from irregularities in the density of the Universe. The problem is the fact that the baryonic matter was ionized until the epoch of decoupling, at a redshift of perhaps 1100. Any irregularities would not have grown before that time, because any attempt by the protons to condense would have been wiped out through scattering by high-energy photons. Condensations of baryons could only form once the Universe was composed of neutral atoms. Even though the Universe has expanded by a factor of about 1100 since then (400 000 years after the big bang), there has not really been enough time for galaxies to have separated so completely from one another if they only began condensing at the time of decoupling.

Neutral dark-matter particles provide one way around this problem. Because they were uncharged, they could have condensed from initial density irregularities well before decoupling. These condensations could have provided gravitational 'wells' into which the baryons fell once they became absorbed into neutral atoms. This mechanism only works if the dark-matter particles have large masses, of the order of a few MeV/c^2 or more. The mass gives them the gravitational attraction that causes the

cold dark matter

condensations, and the large mass means that they have relatively small speeds, also making it easier to form condensations. Because of the low speeds such particles must have had at decoupling, astrophysicists call them *cold dark matter*.

Cold dark matter has been a key feature of most recent attempts to devise detailed theories of galaxy formation and particularly of the way that galaxies cluster together. By simulating this process on computers, astronomers hope to learn how much dark matter is required to produce clusters with the size, shape and spacing that we observe today. These simulations show what sort of irregularities had to be present at the time of decoupling for the present clusters to have formed. These irregularities would have caused slight irregularities in the temperature of the cosmic microwave background. The COBE and WMAP observations of these irregularities are in good agreement with this idea. However, there are quantitative differences. Simulations that use enough cold dark matter to reach the critical density, ρ^c, do form clusters of galaxies, but their distribution is somewhat more irregular and inhomogeneous than the distribution we observe. Simulations that combine cold dark matter and dark energy produce more realistic distributions.

hot dark matter

Another way to smooth out the distribution of clusters is to replace some of the cold dark matter with *hot dark matter*, by which we mean matter that was still moving too fast at the time of decoupling to have provided much of a seed for baryon condensation, at least on the scale of galaxy clusters. Neutrinos with masses of the order of $10\,\mathrm{eV}/c^2$ were once prime hot dark-matter candidates. By now they would be cold, and concentrated in galaxy clusters, contributing to their dark matter. But earlier they would have moved about more energetically, smoothing out the irregularities when the clusters were forming. However, $10\,\mathrm{eV}/c^2$ now appears an unacceptably high value for the neutrino mass, so the notion of mixing hot and cold dark matter now looks rather unattractive. In addition, the WMAP results indicate that fast-moving neutrinos have not played a significant role in the evolution of structure in the Universe. If they had, they would have prevented the early clumping of gas and delayed the emergence of the first stars to an unacceptable extent.

Although neither cold dark matter alone, nor a combination of hot and cold dark matter, seems able to account for the kind of clustering we actually observe, astronomers are not short of other ideas. Alternative ways of forming the appropriate structures have been proposed that make use of 'seed' structures known as 'cosmic strings' and 'cosmic textures'. However, a different approach that does seem to work is to include the effect of a cosmological constant, or equivalently, a constant dark energy and an associated negative pressure. This brings us to our next topic.

5.4   The nature of dark energy

To end this discussion of the density of the Universe and its composition we consider the nature of dark energy. On the basis of a number of lines of evidence, dark energy is currently thought to account for 70% of the total energy density, yet its nature is still a mystery and even its existence remains controversial.

In this section, we shall consider three proposals concerning the nature of the dark energy. First, the possibility that it really should be regarded as a *cosmological constant* of the kind that Einstein introduced in 1917. Second, the suggestion that it represents the energy of empty space (*vacuum energy*), a concept rooted in quantum physics. Third, the notion that it might be the energy of some kind of negative pressure field usually referred to as *quintessence*. We shall consider these three possibilities in turn.

5.4.1 Dark energy as a cosmological constant

Einstein introduced the cosmological constant Λ in 1917. At the time he was trying to produce a 'static' cosmological model, in which there was neither expansion nor contraction. (His goal is now seen as misguided, but the usefulness of the cosmological constant remains open to debate.)

In seeking a static universe, Einstein did not introduce a dark energy to counteract the self-gravitating effect of matter. Rather, he chose to modify the field equations themselves. You saw in Section 2.1 of this unit that the Einstein field equations can be written schematically as

$$\boxed{\begin{array}{l}\text{Ricci curvature of}\\\text{spacetime at any}\\\text{given event}\end{array}} = \boxed{\begin{array}{l}\text{Ten-component source}\\\text{term at the same event}\end{array}} .$$

Using the same schematic notation, we can represent Einstein's modified field equations by

$$\boxed{\begin{array}{l}\text{Ricci curvature of}\\\text{spacetime at any}\\\text{given event}\end{array}} -\Lambda g_{\mu\nu} = \boxed{\begin{array}{l}\text{Ten-component source}\\\text{term at the same event}\end{array}}$$

where $g_{\mu\nu}$ is the *metric tensor*, a quantity already implicitly present elsewhere in the field equations, the ten distinct components of which are the metric coefficients at a given event.

The Friedmann equations that follow from the simpler, original field equations in the case of a homogeneous and isotropic universe, may be represented by Equations 1 and 5 of Section 2.2.

$$\frac{1}{R}\frac{\mathrm{d}^2 R}{\mathrm{d}t^2} = -\frac{4\pi G}{3c^2}(\rho + 3p) \tag{1}$$

$$\frac{1}{R^2}\left(\frac{\mathrm{d}R}{\mathrm{d}t}\right)^2 + \frac{kc^2}{R^2} = \frac{8\pi G}{3c^2}\rho \tag{5}$$

where R, p and ρ are all dependent on time. The analogous equations that follow from the modified Einstein field equations may be written

$$\frac{1}{R}\frac{\mathrm{d}^2 R}{\mathrm{d}t^2} = -\frac{4\pi G}{3c^2}(\rho + 3p) + \frac{\Lambda}{3}$$

$$\frac{1}{R^2}\left(\frac{\mathrm{d}R}{\mathrm{d}t}\right)^2 + \frac{kc^2}{R^2} = \frac{8\pi G}{3c^2}\rho + \frac{\Lambda}{3}.$$

Comparing the corresponding members of the two pairs of equations, it can be seen that the second pair become identical to the first pair if we take the following steps:

1 Let

$$\frac{8\pi G\rho_\Lambda}{3c^2} = \frac{\Lambda}{3} \qquad \text{and} \qquad p_\Lambda = -\rho_\Lambda.$$

2 Use the fact that $\rho_\Lambda = -\frac{1}{2}(\rho_\Lambda + 3p_\Lambda)$ to rewrite the second of the two pairs of equations

$$\frac{1}{R}\frac{d^2R}{dt^2} = -\frac{4\pi G}{3c^2}(\rho + 3p) - \frac{4\pi G}{3c^2}(\rho_\Lambda + 3p_\Lambda)$$

$$\frac{1}{R^2}\left(\frac{dR}{dt}\right)^2 + \frac{kc^2}{R^2} = \frac{8\pi G}{3c^2}\rho + \frac{8\pi G}{3c^2}\rho_\Lambda.$$

3 Rearrange this pair of equations to read

$$\frac{1}{R}\frac{d^2R}{dt^2} = -\frac{4\pi G}{3c^2}\left(\rho + \rho_\Lambda + 3(p + p_\Lambda)\right)$$

$$\frac{1}{R^2}\left(\frac{dR}{dt}\right)^2 + \frac{kc^2}{R^2} = \frac{8\pi G}{3c^2}(\rho + \rho_\Lambda).$$

Now, this last pair is identical to the first pair of equations (Equations 1 and 5) provided we recognize that in those original equations ρ *includes* a dark-energy contribution ρ_Λ and p *includes* an associated pressure $p_\Lambda = -\rho_\Lambda$. This is what we have been assuming all along.

What this establishes is that if we follow Einstein's original approach, there is no need to provide any physical explanation of the dark energy. Dark energy is simply a way of representing the geometrical effect of including a cosmological constant on the left-hand side of the field equations, in terms of contributions to the source terms on the right-hand side.

Physicists, with their taste for interpreting the terms in an equation, generally dislike this approach. Still, from a mathematical point of view we can, superficially at least, say that dark energy has no deeper physical significance according to this interpretation.

5.4.2 *Dark energy as a vacuum energy*

A more palatable approach for many physicists is to think of the dark energy and its associated pressure as properties of the vacuum that arise as a possible consequence of quantum physics. This is so different from the way we normally think about the vacuum that we shall need to consider it in some detail.

Normally, we think of energy as being associated only with particles (including photons). The vacuum, on the other hand, is usually regarded as the absence of everything, so it would seem natural to think it has zero energy. But when we look at things from the perspective of quantum physics, there are new possibilities.

In quantum physics, there can be energy without any particles at all. This is because of the Heisenberg uncertainty principle. According to this principle, it is impossible to define the energy of any quantum state *exactly*. This applies even to the vacuum — the quantum state in which there are no particles at all. It cannot have exactly zero energy because if it did, we would know its exact energy. The energy of the vacuum must be able to fluctuate, so yielding an *average* energy of the vacuum. In fact, this 'busy' nature of the vacuum leaves its mark in the fine details of atomic spectra and elsewhere, so its reality can be confirmed by experiments.

The uncertainty principle says that a quantum state that exists for only a short time Δt cannot have a perfectly-defined energy; the energy uncertainty ΔE must be at least as large as $\hbar/\Delta t$, where $\hbar = h/2\pi$, h being Planck's constant. In the theory of quantum electromagnetic fields (which is called *quantum electrodynamics*), this has a simple but striking consequence. Suppose we consider the vacuum state. We might expect it to have zero energy. But, during an interval of time Δt, we cannot define its energy to be exactly zero: it could have any energy up to $\hbar/\Delta t$. It is therefore possible for photons, for example, to exist with this energy, provided they go away again within a time Δt. These are called vacuum fluctuations, and quantum field theory predicts that a vacuum is in fact always undergoing such fluctuations.

Now, the energy of a photon is hf, where f is its frequency. If we set this to the maximum allowed fluctuation energy, we find that $hf = \hbar/\Delta t$, i.e. $f = 1/2\pi\,\Delta t$. The period of a wave with frequency f is just $1/f$, so Δt must be less than the period of the fluctuating photon. This leads to a simple statement about vacuum fluctuations: a photon of frequency f is allowed to arise spontaneously provided it goes away again after less than one period of oscillation of its electromagnetic fields. The vacuum is full of such photons. Sometimes they are there, and sometimes not, with the result that there is an average fluctuating energy. This is called the *zero-point energy* of the vacuum for photons of frequency f. By adding together the zero-point energy associated with all possible photon frequencies in a given volume of space and then dividing the result by that volume we will obtain an average *vacuum energy* density ρ_v.

zero-point energy

vacuum energy

This vacuum energy must arise in a manner that is independent of the observer, since the theories of physics do not allow us to single out preferred observers. If we make a Lorentz transformation we should expect to get the *same* vacuum energy density. Normally a Lorentz transformation does not leave energy unchanged, and it is even worse with energy *density*, since it does not leave volumes unchanged either. But there is one special case where the energy density is independent of the observer, and that is when there is also a pressure p which is the *negative* of the energy density ρ:

$$p = -\rho.$$

This is just what relativistic calculations of the vacuum energy give. An energy density ρ_v and an associated negative pressure $p_v = -\rho_v$.

A negative pressure is not necessarily unphysical. Whereas positive pressure pushes, a negative pressure pulls. It is usually called *tension*. The quantum vacuum would exert its tension in all directions equally.

Notice that this tension would not be felt as a local force. This is because pressures act only through pressure *differences*. A helium balloon rises upwards in air only because air pressure underneath it is slightly greater than the pressure above it. This results in an unbalanced buoyancy force that, for the helium balloon, is bigger than its weight, so it floats upwards. The 'quantum vacuum' has a uniform pressure, so we do not notice it locally.

Although there are no local forces due to the vacuum pressure, it does have a remarkable effect. We can see this by first of all considering a cylinder fitted with a piston and filled with a gas exerting a normal positive pressure. That pressure can force back the piston, allowing the gas to expand. As it does so, the gas does work on the piston, with the result that the total energy of the gas is reduced — it cools. Since the volume has expanded, the total energy *density* of the gas is reduced also.

Now suppose the cylinder were filled, not with ordinary gas, but with the (quantum) vacuum. With its negative pressure the vacuum would not of itself show any tendency to expand. In order for the cylinder to expand, work would have to be done *on* the vacuum by whatever lay on the other side of the piston. Since the vacuum in the cylinder pulls inwards, yet the piston is being forced in the opposite direction by the outside agency, the vacuum would *gain* energy.

As is well known, the amount of work done (the energy expended) when the point of application of a force, F, moves through a distance, ΔL, is the product $F \Delta L$. If the force is in the form of a pressure p, acting over an area ΔA, then $F = p \Delta A$. Thus, the work done is $p \Delta A \Delta L$ or $p \Delta V$, where ΔV is the change in volume. Since numerically the pressure equals the energy density in this case, the work done by the experimenter is $p \Delta V$. This work increases the energy of the vacuum inside the cylinder, so its total energy is now $\rho(V + \Delta V)$. Since its volume is $V + \Delta V$, its new energy density is ρ, just as before. The expansion of the cylinder did not dilute the energy density!

For cosmology, this is very important. Not only is the vacuum energy density independent of the observer, it is also independent of time: as the Universe expands, the work done by the expansion should continually add to the vacuum energy, keeping its density constant. Thus an energy density obtained from vacuum fluctuations is a universal constant of nature!

Now this constant vacuum energy density ρ_v with its associated negative pressure $p_v = -\rho_v$ seems to have all the properties we require of the dark energy. However there is a major obstacle that prevents us from simply saying the two are the same and writing $\rho_\Lambda = \rho_v$. The difficulty concerns the value of ρ_v.

The total vacuum energy density has a finite value because there is a shortest possible fluctuation time Δt, and hence a largest possible energy $\hbar/\Delta t$. This limit comes from gravity itself: a photon cannot have arbitrarily high energy, because eventually the energy in one wavelength would be so large that its self-gravitation would cause it to form a black hole. As we saw in Unit 12, black holes form when the total mass M is contained in a region of radius smaller than about GM/c^2. (Actually $2GM/c^2$, but for present purposes we can ignore the factor 2.) Setting $M = \Delta E/c^2$, and using $\Delta E = \hbar/\Delta t = \hbar c/r_{Pl}$, where r_{Pl} is the distance travelled by light in time Δt. Taking this to be the radius of the black hole, we obtain the condition that the wavelength, λ, associated with the photon is such that

$$\lambda > r_{Pl} = \left(\frac{G\hbar}{c^3}\right)^{1/2} \sim 1.6 \times 10^{-35} \, \text{m}.$$

The special length r_{Pl} is defined only by fundamental physical constants, and is called the *Planck length*. The mass of this black hole is the *Planck mass*, $m_{Pl} = (c\hbar/G)^{1/2} \sim 2 \times 10^{-8}$ kg.

Planck length

Planck mass

Objective 6

Planck time

SAQ 21 (a) Go through the calculation indicated above and derive the Planck limit on the wavelength of photons.

(b) There is a third fundamental unit, the *Planck time*, defined as the time it takes light to travel a Planck length. Show that the Planck time is $t_{Pl} = (G\hbar/c^5)^{1/2}$, and calculate its value.

Handwritten margin notes:

pressure $= Pa = Nm^{-2}$

energy density $= Jm^{-3} = Nm\,m^{-3}$,

so pressure & energy density $= Nm^{-2}$

$M = \Delta E/c^2$ where $\Delta E = \hbar c/r_{Pl}$

so $M = \hbar/r_{Pl}c \, \& \, r_{Pl} = \left(\frac{G\hbar}{c^2 r_{Pl}}\right)$

so $r_{Pl} = \left(\frac{G\hbar}{c^3}\right)^{1/2}$.

(c) Express the mass of the proton in units of the Planck mass. Express the size of the proton (10^{-15} m) in units of the Planck length. Express the lifetime of the most unstable known particles (about 10^{-20} s) in units of the Planck time.

The Planck length, mass, and time are remarkable quantities, because they depend only on fundamental constants of nature, not on the properties of any known particles. They seem to give us natural units for physics. But, as SAQ 21 showed, when we explore physics in the laboratory, we are in fact far from the Planck regime. Nature has, for example, a natural length-scale r_{Pl}, and we could, in principle, use this to measure other distances, rather than the arbitrary human-defined unit of the metre. But this scale is inconveniently small, and will not prove useful unless we deal with physics that occurs on this scale. Since it is built from G, c, and $\hbar$, physicists believe that it will be relevant in physical situations in which gravity, relativity, and quantum mechanics are all important. This would be a theory of quantum gravity — when we finally have a theory that predicts quantum effects in the gravitational field, we expect these effects to be important on the Planck scale.

Planck density

Planck black hole

Finite though it may be, the Planck length is still a tiny wavelength, and therefore the 'uncertainty energy' associated with it is huge. The biggest energy fluctuation that can occur in a region of size r_{Pl} is $m_{Pl}c^2$. The energy density this creates is obtained by dividing by the volume r_{Pl}^3. Physicists call it the *Planck density*, and it is the energy density of the *Planck black hole*, about $\rho_{Pl} = c^7/G^2\hbar \sim 10^{114}$ J m^{-3}. Now this 'biggest energy fluctuation' will set the scale for the total energy density of the vacuum. So, ignoring all the details, we expect $\rho_v \simeq 10^{114}$ J m^{-3}.

Now, this is a huge energy density, much too great to go unnoticed locally and also too great to be the dark energy density ρ_Λ. The latter, as we have seen, is thought to be about 70% of ρ_0^c, and we saw in Section 3.2 that $\rho_0^c \approx 10^{-9}$ J m^{-3}. Thus, our estimate for the vacuum energy ρ_v exceeds ρ_Λ by a factor of more than 10^{120}.

The failure of such a promising candidate as ρ_v to match the value of ρ_Λ is deeply disappointing. But the overall situation is much worse than disappointing. The existence of the vacuum energy is required by local particle physics, independently of any cosmological argument. So the prediction of such an unacceptably high value of ρ_v is a disaster in its own right. It directly conflicts with everyday laboratory physics.

Currently, there is no accepted solution to this particular problem though there are plenty of speculations. Many physicists hope that as our understanding of quantum physics improves a way will be found to 'tame' the quantum fluctuations, so that an acceptable value of ρ_v will emerge. Being optimistic, it might even be hoped that this 'acceptable' value of ρ_v will turn out to be just that needed to account for ρ_Λ. However, this has not yet been accomplished and it remains a hope rather than an expectation.

5.4.3   Dark energy as quintessence

A third way of accounting for dark energy is by supposing it to be the energy associated with some new kind of matter or, perhaps more realistically, some previously undetected field that fills all of space and has a positive energy density but a negative pressure. One specific proposal regarding this new cosmic ingredient dubs it '*quintessence*' (after the fifth of the ancient Greek 'elements').

Quintessence is more general than our earlier proposals in the sense that its energy density ρ_q and pressure p_q are related by an *equation of state* that takes the form

$$p_q = w\rho_q$$

If $w = -1$, then quintessence would have the same large-scale effect as the cosmological constant. But if w had some other value, not too far from -1, then quintessence might be consistent with the existing cosmological observations while also holding out the possibility of new phenomena that could be sought.

The evidence regarding quintessence is slight, but not promising. According to the WMAP results, $w < -0.78$ with 95% confidence. Thus, quintessence is not ruled out, but it will, at best, be hard to distinguish its effects from those of a cosmological constant.

6 Inflation

Study comment

Please refer to the study comment at the beginning of this Unit concerning study strategy if you have fallen behind.

The observational evidence reviewed in Section 5 suggests that the total energy density of the Universe is very close to the critical density. Computer simulations tell us that the structure we see could be consistent with a Universe that is close to the critical density, provided it has the right mix of constituents.

In one sense, having a density that is close to critical is philosophically appealing to many scientists. But in another sense, it is hard to understand. To see why, suppose the Universe is matter-dominated and has a density that today is half the critical density. The Universe would then expand forever and, as it does so, the difference between its true density and the critical density at any time continuously increases. The following SAQ explores how this comes about.

Objectives 2 and 8   **SAQ 22**   Consider a matter-dominated universe with $k = -1$.

(a) Use Equation 5 to show that the rate of expansion eventually becomes a constant.

(b) As R becomes large, how does the Hubble parameter depend on R?

(c) The critical density, ρ^c, at any arbitrary time t, is given by Equation 7. How does ρ^c vary for large R? How does the ratio ρ/ρ^c vary for large R?

What we discover from this SAQ is that the ratio ρ/ρ^c, of the density at time t of a matter-dominated universe to the corresponding value of the critical density at that time, approaches zero for large R. In fact, if in the initial stages of the big bang the density ρ had been only slightly less than the critical density at that time, the two would have rapidly diverged — to the extent that to find the density today still within a factor of 2 of the critical value is nothing less than astonishing. Indeed, to get a density today that is as close to critical as it appears to be, the density would have to have been within *one part in 10^{15} of the critical density at the time of nucleosynthesis.*

This immediately raises the question: how did they get to be so exceedingly close? Many physicists take this coincidence to mean that there must be some reason — some mechanism — for *ensuring* that the difference between the critical density and the actual density is zero. If the Universe had *exactly* the critical density in its early stages, then it would have exactly the critical density for all time.

inflation

One of the great attractions of the theory of *inflation* we are about to describe, is that it predicts a density today that would be essentially critical. Inflation is a period of exponentially rapid expansion in the very early universe, well before the epoch of nucleosynthesis. We explain in this section how it may have happened, and how it would explain a number of puzzling facts.

Inflation is not as yet an established physical fact. Rather, it is a phenomenon that could have occurred if the theory of high-energy physics has a suitable form. Since we do not know the correct laws of physics at very high energies, we cannot be sure that they would have allowed inflation to happen. However, it seems to be a common feature of a large class of theories, and it has so many attractive properties, that it has generated a great deal of discussion. Some of its predictions are testable, and by testing them astronomers will be shedding light not only on cosmology, but also on the fundamental laws of physics at high energy.

6.1   What inflation tries to explain

Before describing what inflation is, it is useful to begin with its motivation. Despite the success of big bang theory in describing the observed Universe, it raises a number of problems. Here is a partial list:

1 Why is the density of the Universe so nearly critical? (We have already explained why this is a difficulty.) This is often referred to as the *flatness problem* since critical density implies $k = 0$ and a flat spatial geometry on the cosmic scale.

flatness problem

2 How did the Universe get to be so homogeneous and isotropic on the large scale? This is a remarkably serious problem. If we look to large distances with our telescopes in any particular direction, we see quasars at redshifts of $z = 1$ or more. That means that the Universe was only half of its present size at the time they emitted the light we are receiving now. It was also therefore no more than about half its present age. Now, if we turn our telescopes around and look in the opposite direction, we again see quasars at redshifts of 1. There are the same numbers of quasars in both directions, and the same sorts of galaxies near them. The Universe looks the same in both directions. Yet the two regions are so far distant that light has taken almost the age of the Universe to reach us positioned as we are halfway between them. It would therefore take twice that time to have passed from one of those regions to the other — a clear impossibility, assuming that light has always travelled with the speed it has today. Thus we appear to be observing two regions of the Universe which have always been independent of each other. So how was it they were formed in such a way as to be almost identical to each other? Until the idea of inflation, astronomers simply had to accept that the homogeneity of the big bang was an unexplained fact. Inflation as we shall see, by contrast, offers an explanation for what astronomers sometimes call the 'homogeneity and isotropy problem', but is more commonly referred to as the *horizon problem*.

horizon problem

This problem shows up in an extreme way in the WMAP observations of the cosmic microwave background radiation. At the time of decoupling, according to big bang theory, two different parts of the Universe could not have communicated with each other if they are separated by more than 5° on the sky today. But WMAP detected significant correlations in the variations in the temperature of the background radiation on angular scales of 10°. How did these correlations arise? Not only that, the background radiation shows an extreme degree of overall isotropy across the entire sky. How did this come about?

3 A third problem is one that has seriously worried particle physicists. It

is the problem that there are no *magnetic monopoles*. Magnetic monopoles are the magnetic analogue of electric charge: a monopole would be an isolated north pole, or an isolated south pole. (An ordinary magnet, which has both a north and a south pole, is called a magnetic *dipole*.) The Universe has plenty of electrically charged particles, but none that have a net non-zero magnetic charge. The only way we appear to be able to generate magnetism is by moving electric charges, and this has the consequence that north poles are always accompanied by south poles. This follows from the equation $\operatorname{div} \mathbf{B} = 0$ (Unit 4), which says that magnetic lines of force are closed: they do not terminate anywhere. By contrast, the analogous equation for the electric field is $\operatorname{div} \mathbf{E} = \rho_e/\epsilon_0$, where ρ_e is the density of electric charge. This acknowledges that isolated electric charges do exist.

Now, there is nothing fundamental about the fact that the divergence of the magnetic field becomes zero. It was included by Maxwell in his equations because isolated magnetic charges were not observed. If experiments were subsequently to demonstrate that they did exist, it would be simple to change the magnetic-field equation to $\operatorname{div} \mathbf{B} \propto \rho_m$, where ρ_m would be the density of magnetic charge. This could be absorbed into Maxwell's equations without inconsistency. Indeed, in quantum field theories in which electromagnetism is unified with other forces of physics, magnetic monopoles do appear naturally. They may have a large mass, and therefore be hard to create today, but in the early Universe there was plenty of energy, so they would have been created in abundance. They should be far more abundant than current limits from experimental searches prove them to be.

The theory of inflation provides answers to each of the three problems outlined above.

6.2   How inflation works

The fundamental idea underlying inflation is the notion of vacuum energy. However, in this case, it is not the vacuum energy of the present-day Universe, but rather that of an early phase of the Universe's existence. We shall call this 'early' vacuum energy ρ_i to indicate its link with inflation.

6.2.1   Inflation: a universe driven by vacuum energy

We already know, from our earlier discussion of vacuum energy (Section 5.4.2), how the Universe might behave if this early vacuum energy ρ_i were present. If ρ_i was the *dominant* contribution to the energy density in some early era, then during that era we would find $\rho \approx \rho_i$ and $p \approx -\rho_i$. As a consequence, the Friedmann equations predict that, during that era, R would evolve according to the equation:

$$\frac{1}{R}\frac{d^2 R}{dt^2} = -\frac{4\pi G}{3c^2}(\rho + 3p)$$

$$= -\frac{4\pi G}{3c^2}(\rho_i - 3\rho_i)$$

$$= \frac{8\pi G}{3c^2}\rho_i. \tag{12}$$

The right-hand side of this is *positive*, so the expansion of the Universe during this early era would have accelerated just as we think it may be accelerating now due to $\rho_{\Lambda,0}$.

The expansion produced by this early vacuum energy grows exponentially. Since the right-hand side of Equation 12 is constant (remember that the vacuum energy density in an expanding region is constant), the acceleration $\mathrm{d}^2 R/\mathrm{d}t^2$ is proportional to R. As R gets larger, the acceleration also gets larger, in accordance with

$$R \propto \mathrm{e}^{t/\tau}, \tag{13}$$

where the constant τ is given by

$$\tau = \left(\frac{3c^2}{8\pi G \rho_i}\right)^{1/2}. \tag{14}$$

inflation

This is the characteristic time-scale for the expansion. This early phase of exponential growth of the Universe is called *inflation*.

Objective 6   **SAQ 23**   Demonstrate by substitution that Equation 13, with the given formula for τ, is a solution to Equation 12.

If ρ_i is equal to the Planck energy density we estimated in Section 5.4.2, then, by setting ρ_i equal to ρ_{Pl}, we find that the exponential growth time is

$$\tau = \left(\frac{3G\hbar}{8\pi c^5}\right)^{1/2} \approx 2 \times 10^{-44} \text{ s}.$$

Inflation driven by the Planck energy density would grow on the Planck time-scale. This would be inevitable, since our formulas so far only have the constants G, c, and $\hbar$ to manipulate, so any time-scale that comes out must be a simple factor times the Planck time.

It is clear that the Universe today is *not* inflating like this. But the theory of cosmological inflation asserts that, at a very early time, the Universe *did* briefly go through a phase of inflation. At the end of that period of superfast expansion, the Universe was left to decelerate to its present, much more leisurely rate of expansion. In order to see why the Universe is not still inflating today and how this might have happened in the past, we must introduce more physics into our model of the Universe. In addition to the early vacuum energy needed to drive inflation, we must somehow find a way of starting inflation and then of stopping it again. It may be that this is mainly a consequence of the Universe's declining temperature, as we shall now explain.

From the point of view of quantum physics, the vacuum is simply the quantum state of lowest energy. Despite the enormous vacuum energy we have talked about, any other state has particles (photons) in it, so it has an even larger energy. Now, the laws of interactions among elementary particles are complicated in that they suggest the lowest energy state may depend on the mean *temperature* of the particles, i.e. on how fast they might be moving relative to one another. At very high temperatures, when the kinetic energies of all the particles are near the Planck energy (and are therefore much, much larger than their mass energies), the vacuum state is very simple. All of the particles behave in essentially the same way. This is how the Universe was, right after the big bang. It is possible that as the Universe expanded and cooled, a different quantum state — with more complicated properties — became the one with lowest energy. This process is illustrated schematically in Figure 15.

The switch from one vacuum state to another is an example of a process called *spontaneous symmetry-breaking*, because the simplicity (symmetry) of the original vacuum is replaced (broken) by the complexity we see in particle physics today.

A more common example of spontaneous symmetry-breaking is the process of crystallization. As a molten material cools, it sometimes forms domains in which the atoms are strongly ordered. The original, molten state was disordered and therefore very symmetrical: it could be rotated and it would look the same. But when a crystal forms, the symmetry is broken and replaced by an ordered state in which some directions are different from others. In most situations, the crystals form in different parts of the material with different orientations, and there are irregular and complicated boundaries between the different domains. This will have an analogy in the behaviour of inflation.

Something similar is thought to have happened in the early Universe. At the high temperatures then prevailing, there was a sea of quarks (the constituent particles from which today's protons and neutrons are made), electrons, neutrinos and photons. They all interacted with one another equally. But as the temperature fell, something happened that caused the state of lowest energy to be one in which quarks and electrons have mass and photons don't. Unlike the process of crystallization, the initial symmetry here was not one of rotations in space; rather, it consisted of an equality, or democracy, among particles. This symmetry was broken when some particles acquired more mass than others, with the result that physics from this time on had less symmetry than it did at the higher temperatures.

Very soon after the big bang, the Universe was so hot that the highly symmetric vacuum state was the relevant one, rather than our present vacuum. The Universe was full of particles, all at first having an energy comparable to the Planck energy. The total energy density consisted of the energy of the particles, plus the uncertainty energy of the original symmetric vacuum state. But as the Universe cooled the symmetry became broken and a new vacuum state of even lower energy became accessible. The Universe, however, was still in the original more highly symmetric state which had become a 'false vacuum' state. The difference in vacuum energy density between the new 'true' vacuum and the old 'false' vacuum would act like a cosmological constant, creating exponential expansion as long as the Universe remained in the false vacuum.

It is believed that this effective cosmological constant's energy density was determined by the energy at which the first spontaneous symmetry-breaking occurred, thought to be roughly 10^{14} GeV per particle. By confining this energy to a volume of the size of the wavelength of a photon of this energy, as we did to get the Planck density, we arrive at a characteristic energy density of 10^{94} J m^{-3}. Large as it is, it is small compared to the 10^{114} J m^{-3}, which was the energy of the particles that had Planck energies at the original high temperatures.

spontaneous symmetry-breaking

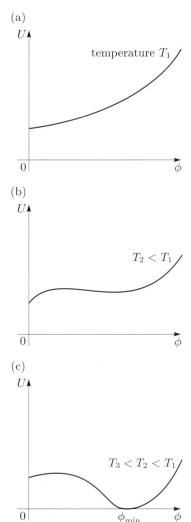

Figure 15   A highly schematic illustration of the development of a new vacuum state, with less energy than the original vacuum, as the Universe cools. The figures depict the changing potential energy associated with some hypothetical field ϕ as the cosmic temperature T decreases. As a new minimum develops the preferred value of ϕ changes from its original (symmetric) value $\phi = 0$, to a (less symmetric) $\phi = \phi_{\min}$.

Objective 6
1 GeV = 10^9 eV.

SAQ 24 Determine the characteristic energy density of the vacuum fluctuations associated with photons of energies of 10^{14} GeV.

Objective 6

SAQ 25 Earlier it was remarked (in connection with SAQ 23) that if the vacuum energy density ρ_v were the Planck density, then we would have exponential growth with a time-scale of $\tau \approx 10^{-44}$ s. We now have a more realistic energy density of 10^{94} J m^{-3}. Refer back to Equation 14, and using this value of the density, evaluate the constant τ.

We see from that last SAQ, that with the vacuum energy density given above, the exponential time-scale was about 10^{-34} seconds. The Universe was roughly doubling its size every time the clock ticked 10^{-34} s!

This could not go on forever, because as the Universe expanded, the material in it cooled further, and the state of lowest energy changed. At random places around the Universe, the particle fields began to adopt the values characteristic of the spontaneously broken state that we see today. The 'true' vacuum began to establish itself as the false one died.

Something remarkable happens when the false vacuum dies. When the transition occurs, the difference in energy between the false and true vacuums must be released, and this heats up the Universe. (The same happens when a crystal forms: the latent heat of crystallization is released.) At the end of inflation, this released energy did not go into re-heating the Universe; rather, it went into creating the particles that acquire mass from the symmetry-breaking. *This is how, in an inflationary model, essentially all the particles we see in the Universe today were formed.* According to this model, most of the contents of the Universe were *not* created instantaneously at $t = 0$. Only a few particles were created at that stage; most came into being a fraction of a second later — through the inflation process.

The negative pressure associated with the energy of the original false vacuum goes away as this energy is converted into real particles; the random motions of these particles produce an ordinary positive pressure. The exponential expansion ceases, and the Universe starts to decelerate. But it does so from the enormous initial expansion speed provided by inflation.

This is the point where the standard big bang picture takes over. When inflation ends, the Universe is a hot gas of particles moving about in the present familiar vacuum. From this point on, it is possible to discuss the subsequent evolution of the Universe in terms of 'time since the big bang'. Now, in fact, we can't know how long it was from the actual instant of the big bang (the true $t = 0$), since we know little about the time the Universe spent in its *pre*-inflationary phase. What is meant by 'time' in most discussions, including ours here, is the time that it would have taken the Universe to reach a certain state from a standard big bang, without inflation. Time in this sense is not really what is ticked by a clock, but just a marker to calibrate the physical condition of the Universe, such as its temperature. But once inflation ends, and the standard big bang takes over, then the differences in time between two different epochs are physical. We may not know how long it took to go from the true big bang to the epoch assigned a time of, say, 10^{-34} s, but the time it took the Universe to go from 10^{-34} s to 10^{-32} s was simply $(10^{-32} - 10^{-34})$ s $= 9.9 \times 10^{-33}$ s.

The stage at which most of the matter of the Universe was created.

6.3   Does inflation solve the problems of the big bang?

It is easy to see how inflation solves the horizon problem (i.e. the homogeneity/isotropy problem). If the period between the onset of inflation and its cessation is long enough, the expansion would have inflated any small region into an enormous size. The Universe we see today could have come from something very small, so small that light could cross it even in the short time between the big bang and the onset of inflation, and this implies that the region would have had time to smooth itself out. Of course, any particles present in that region before inflation would have been dispersed to enormous separations. We would not expect to see those particles today because their density has become so dilute. The ones we do see (and are made of) were created at the end of inflation, filling in the vast, smooth, empty region that the original domain became. In this picture, the distant quasars and the various regions of the Universe at decoupling were all part of the same original tiny domain.

This also shows how inflation solves the monopole problem. If inflation occurs after monopoles formed, they too will be dispersed so far apart that the chance of our encountering one now would be infinitesimal. Provided the temperature at inflation is smaller than that needed to create monopoles (which is true for present models), the end of inflation would not have seen the creation of any more of them.

Inflation also solves the problem of why the total energy density is so close to critical: the process by which particles are produced during inflation occurs at just the right rate required to make the density exactly critical. The exponential expansion phase wiped out any memory of the initial expansion velocity before inflation set in, leaving the conventional Universe with just the right expansion rate. It follows that at the end of inflation the scale factor was given to high accuracy by the relation $R = Ae^{t/\tau}$, with τ as given above (see Equation 14). The Hubble expansion rate is $(\mathrm{d}R/\mathrm{d}t)/R$, and for this expansion law we find $H = 1/\tau$. The square of the Hubble expansion rate is then

$$H^2 = \frac{1}{\tau^2} = \frac{8\pi G\rho_\mathrm{i}}{3c^2},$$

where here ρ_i is the energy density driving inflation. We have seen that, as inflation ends, this energy remains in the particles, while the inflationary (negative) pressure changes. Therefore, the subsequent Hubble expansion starts with a Hubble expansion rate H and energy density ρ related by

$$\rho = \frac{3H^2c^2}{8\pi G}$$

which is exactly the relation that the expansion rate and density satisfy for a Universe with a critical density. The Universe was critical to a very high accuracy when the inflationary epoch ended, and it will therefore have stayed critical ever since. This solves the flatness problem.

Inflation also explains things we did not ask it to explain. Chief among these is why the Universe is expanding at all: why was the big bang so big? Inflation assumes only that, at the very earliest times, there was a region of the Universe that was expanding, however slowly, just enough to cool its contents to the point where the effective cosmological constant took over. After that, inflation gives the Universe its enormous push, and it is set into an expansion that will last far longer than 10^{10} years. It is possible to make inflation work even if the initial conditions before inflation were highly variable from one place to another.

Another bonus of inflation is that it provides an initial spectrum of density irregularities and gravitational waves at an early time that can lead to observable consequences. Studies of the distribution of galaxies and clusters, of the irregularities in the cosmic microwave background, and (in the future) of the gravitational waves left over from this time all provide constraints on how long inflation lasted, how quickly or slowly it ended, and where the energy of the false vacuum was deposited.

So, should we take inflation seriously? Can we really believe any physical idea that deals with the first 10^{-32} s? Is the early vacuum energy density ρ_i plausible, or just a strange creation of an incomplete theory of physics?

It must be stressed that inflation is a physical idea, not a philosophical one, and it will stand or fall by its predictions. One can perform many calculations about the details, such as how much matter is created at the end of inflation and how much irregularity one expects in the mass distribution of the Universe after inflation finishes. The earliest version of inflation, proposed by the physicist Alan Guth, was discarded after it was shown to produce too much density irregularity today. Its replacement, called 'new inflation' and the many other variants that have been proposed since, are going through many tests and refinements.

Objective 8 **SAQ 26** Imagine a balloon being blown up (inflated) from being very small to being enormous. Consider the way the curvature of a tiny area of the balloon changes. Why can this be used as a good analogy for the process of inflation we have been considering?

Objective 8 **SAQ 27** According to the inflation idea, the density of the Universe should end up critical, and the geometry of the Universe should be flat. Does that mean the curvature parameter k, appearing in the Robertson–Walker metric, takes on the value zero after the inflation period?

6.4 Cosmic defects

Inflation is only the most spectacular example of what can happen if modern theories of particle physics are applied to the early history of the Universe. Even if inflation never occurred, we might still find the Universe filled with what are called 'defects', primarily *cosmic strings* and *cosmic textures*. These also arise from spontaneous symmetry-breaking.

When a symmetry breaks spontaneously, it breaks randomly. In different parts of the Universe, the values of certain fields in the theory will have random aspects, just as the different crystal domains in a freezing liquid will have different, unrelated orientations. In particle physics, the randomness is in values of 'gauge fields', some parts of which are arbitrary.

If you have already studied an advanced course on electromagnetism, you might like to note that the vector potential of electromagnetism is an example of a gauge field because it can be changed in certain ways without changing the electric and magnetic fields it describes.

But when different regions of space have different values of these fields, then real physical effects can result from the *changes* in the gauge fields from one place to another: these changes are usually observable. So it could happen that, at the time the symmetry breaks, one part of the Universe acquires a particular value of some gauge field, and another part acquires a different value. These parts could be so widely separated at this point that they could not coordinate the values they have for the gauge field. But as the Universe develops, the regions grow and so come into contact with each other — in much the same way as the various crystals that form throughout a supercooled liquid grow until they come into contact with each other. It is then across the boundary between one region and its neighbouring region that the differences become noticeable.

Spontaneous symmetry-breaking can have occurred many times since the big bang. The epoch of inflation was (if it happened) a spectacular one. But it could have happened to other fields at lower temperatures later on, and several times. It need not have led to inflation: the details of the way the breaking occurs are crucial to whether there is inflation or just some local effects, which we call defects.

In some kinds of theories, these differences lead to cosmic strings, which are long thin concentrations of energy trapped by the differences in the gauge fields around them. Inside the string, space is still in the old 'false' vacuum, which has lots of energy relative to the true vacuum. If the gauge field is associated with the unification of the strong, electromagnetic, and weak interactions, then the energy can be enormous: the strings can have a mass per unit length of 10^{21} kg m^{-1}. Such isolated strings can serve as seeds for galaxy formation, since they exert significant gravitational attraction on surrounding matter. They can also form loops and decay, emitting gravitational radiation to form an incoherent cosmological background. This is one of the ways that astronomers will be searching for them in the coming years.

In other theories, the gauge fields form cosmic textures, which are three-dimensional knots of energy. As the Universe expands, these decay with bursts of radiation, and these could push the ordinary matter around, leading to condensations that form galaxies. The distribution of the galaxies would be different from that which would arise if strings were the seeds, and this may be one way to distinguish between these models.

Ultimately, the best way to find these defects is in the irregularities they produce in the cosmic microwave background temperature. Recent observations have given little support to these ideas, and with more sensitive experiments, we may be able to rule them out completely.

7 A brief history of the Universe

We are now in a position to draw all the threads together and set before you our best attempt to explain the evolutionary history of the Universe.

7.1   The instant of the big bang: speculation

It is thought that the instant of the big bang, $t = 0$, marked the beginning of the Universe — its formation out of nothing. The early contents of the observable Universe were concentrated at a point of infinite density. But not only the contents. The entire space of the observable Universe also began as a point of no size. In other words, we do not have in mind a pre-existing void in which, at some particular point, the contents of the Universe suddenly appeared, and from which they are now spreading outwards to fill the rest of the surrounding empty space. Initially there was neither space nor time, they came into existence as a result of the big bang.

creation of space

This idea calls for a radical change to our normal thinking. Take for example the question often asked: 'What caused the big bang?' It presupposes a cause followed by an effect — the effect being the big bang. But such a causal chain must be one existing *in* time: cause followed in time by effect. But we are now saying there was no time before the 'effect' — so how could there have been a 'cause'? Questions concerned with what caused the big bang may well be meaningless, and therefore unanswerable.

All of this assumes of course that there actually *was* a $t = 0$. According to an idea advanced by Stephen Hawking, time might not have had a clearly defined beginning. As one imagines going backwards in time towards the supposed $t = 0$, the character of time itself could change — it could become progressively more and more like the three spatial dimensions. Whether there is anything in this idea or not, the broad conclusion would be the same: there was no time before the big bang.

Clearly there is no way our physical theories can deal with conditions of infinite density concentrated at a point. The instant of the big bang (assuming there was one) lies beyond our current understanding. But not only that, we do not believe that Einstein's theory of relativity should in any case be valid at arbitrarily small times. The reason is that it does not incorporate quantum effects into gravity itself.

7.2   The first microsecond: speculation and new theories

As we mentioned earlier, one of the most pressing theoretical problems in physics is to devise a quantum theory of gravity. At present, theories that are called 'supersymmetric' seem good candidates. But whatever the ultimate theory, it seems clear that the Planck length and density will play key roles, since they depend only on fundamental constants of nature. Cosmologists therefore expect that the nature of the Universe will be completely different from the picture given by general relativity before (if that means anything) it cooled to the Planck temperature of 10^{31} K, which is the temperature at which the typical energy is c^2 times the Planck mass.

So, it is not clear that we can talk about the age of the Universe in a sensible way this early in its history. To associate an age (say, the Planck time of 10^{-43} s) with this temperature we have to assume that the dynamics are described by the standard big bang, which may not be the case. Therefore, it is more correct to mark the expansion of the Universe by its decreasing *temperature*, and to leave it to future theories to work out how the temperature changes with time.

After the Planck temperature, when the Universe had the Planck density of 10^{114} J m^{-3}, we might hope to describe the evolution with the correct high-energy physics theory, if we had one that we believed. This might predict a rather gentle evolution. During this period, any density inhomogeneities or temperature differences would be smoothed out by frequent collisions between the particles. This process of thermalization took place throughout the volume that would later expand and become our present observable Universe. The gentle expansion continued until the temperature fell to the symmetry-breaking energy, about 10^{14} GeV, or 10^{27} K. This period is represented by the first section of the curve shown in Figure 16 — the part labelled 'thermalization'.

Planck density p45

Study Note: The remainder of this sub-section, together with Sections 7.3 to 7.6, essentially consist of a commentary upon the main features of the curve contained in Figure 16. You should keep referring back to this all-important figure.

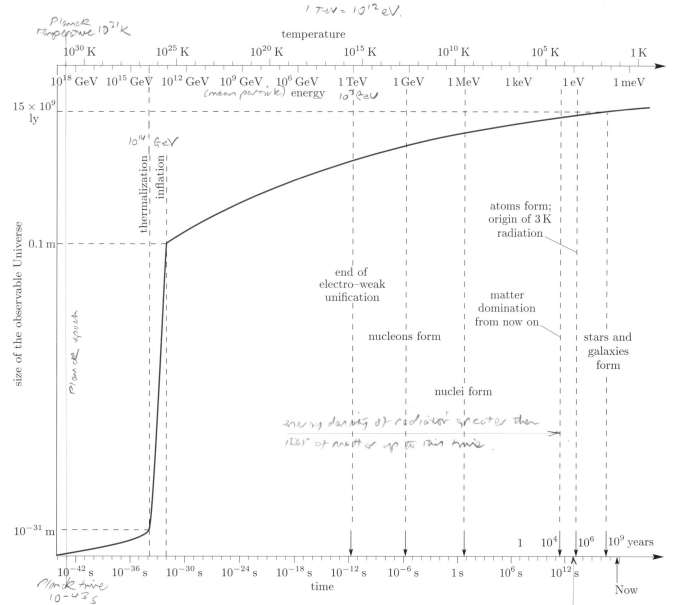

Figure 16   A summary of the development of the Universe as a function of time and temperature.

1 GeV = 1,6×10⁻¹⁰ J

Comment on Figure 16, non-assessable. You might notice in Figure 16 that the energy E (and temperature, T) scales are linked to the time t scale. The relationship is $E \propto t^{-1/2}$, as you can see from the scales. This is because $E_{photon} \propto 1/\lambda_{photon} \propto 1/R$ and *in a radiation-dominated* F–R–W universe $R \propto t^{1/2}$. ($\lambda \propto R$ since the photons expand with the scale factor of the Universe.) However, in a matter-dominated universe $R \propto t^{2/3}$ so there is a subtle change in the relationship at the beginning of the matter-dominated era.

According to present theories, spontaneous symmetry-breaking began to differentiate the strong force from the weak and electromagnetic interactions, and initiated inflation. After that, the Universe expanded dramatically and exponentially, until symmetry-breaking finished. Here we have a prediction of the dynamics of the expanding Universe. The end of inflation should certainly have occurred by 10^{-32} s after it started, and so we say that at this point the Universe was 10^{-32} s old. From now on, we can use time, temperature, or energy as useful markers of the evolution of the Universe. We shall in fact use all of them in our discussion below.

After this time, the Universe is probably described well by the conventional homogeneous isotropic big bang model. There would be further symmetry breaking. For example, after the strong force separated out, the electromagnetic force and the weak force (which is responsible for certain types of radioactivity) continued to be indistinguishable for a while — a combined force known as the electro-weak interaction. Then at an energy of 10^3 GeV (or a time of 10^{-12} s), they too separated out to become distinct interactions. There was the possible formation of cosmic strings, and the annihilation of matter and antimatter to leave a slight excess of matter, from which everything we now see is made. But the gravitational arena in which all this happened was the standard Friedmann–Robertson–Walker Universe.

7.3   The formation of conventional matter and nuclei: from 1 microsecond to 1 hour

By the time the Universe was 1 microsecond old, the temperature had fallen to 10^{13} K, so the mean particle energy was about 1 GeV. This is the energy equivalent of the mass of a proton. At this point, many of the more exotic particles had largely disappeared, and the quarks had combined to form protons and neutrons. There remained also electrons, neutrinos, and photons — and probably also the dark matter that exists today. There were about 10^9 photons per nucleon, and similar numbers of neutrinos. There were roughly equal numbers of neutrons and protons, and (for charge balance) of electrons. From this point on, the evolution was governed by laws of physics that are well-understood. Initially neutrons were interacting and changing into protons, and at about the same rate, protons were changing back into neutrons (hence the proton/neutron ratio was about unity). But below a temperature of 10^{11} K, the energy of the interactions was reduced to a level at which the excess mass of the neutron compared to that of the proton inhibited the reverse reaction. From here on the protons began to dominate in number, yielding eventually a ratio of 87% protons to 13% neutrons.

At 1 microsecond there were still no nuclei as we understand them today. The energies required to split up nuclei are typically in the MeV range, so no nuclei could survive in the Universe until the energies of the particles had dropped to this level. This occurred at a temperature of 10^9 K, and a time of the order of 1 minute. At that stage, deuterium and helium began to form, together with traces of other light elements. These processes, together with the manner in which the free neutrons (the ones that failed to get incorporated into nuclei) then decayed leaving protons and electrons, were described in Unit 13. All of this was finished by the time the temperature fell to 5×10^8 K, when the Universe was about 10 minutes old.

Fig 28 of unit 13 p48

7.4 Radiation, with a dash of matter: from 1 hour to 380 000 years = decoupling epoch.

So, by the end of the first hour, fusion reactions had fizzled out. During the next 10 000 years, the energy density of the radiation continued to be greater than the energy density of matter. It is quite difficult to imagine such a situation because nowadays the opposite is the case. For instance in the Solar System, most of the energy is locked up in the masses of the Sun and the planets. But in the early days of the Universe, the situation was very different — the energy density of the radiation was far greater than that of matter, and the Universe was radiation-dominated rather than matter-dominated.

It is easy to see why such a change-over has since taken place. As we explained in Section 2.2, the energy density of matter is proportional to $1/R^3$, while the energy density of radiation is proportional to $1/R^4$. As time passed, the Universe expanded and R increased. The energy density of both matter and radiation decreased, but the contribution from matter decreased less rapidly than that from radiation. About 10 000 years after the big bang, the two forms of energy contributed equal amounts to the total energy density. From then on, the density of radiation fell well below that of matter.

The next 380 000 years or so were rather boring — nothing was happening in the expanding Universe except cooling. The nuclei and electrons were still separate, in the form of matter called plasma. Unlike a gas, which is made up of neutral atoms, such a plasma is practically opaque to light and other radiation because the charged particles, especially the electrons, scatter radiation strongly. So not only was there nothing interesting to see, one could not have seen it even if there was!

7.5 Amorphous matter: from 380 000 to 10 million years

By the end of the first 380 000 years another, more dramatic change had occurred — the appearance of the first neutral *atoms*. This is the time of decoupling — the origin of the cosmic microwave background radiation.

To see how neutral atoms condensed out of the plasma, we need another characteristic energy: the ionization energy of hydrogen. This is 13.6 eV. Once the typical energy of the photons fell below this, neutral hydrogen could begin to form. However, it had a hard time at first. The reason was the huge imbalance between the number of photons and the number of nucleons. With 10^9 photons for each nucleus, as long as only a small fraction of the photons had energies above 13.6 eV they were able to keep the matter ionized. In any gas, there is a random spread of energies. So it was not until the *average* energy of the photons fell to about 0.25 eV that there were too few photons above 13.6 eV to keep the nuclei ionized. This is the epoch of *decoupling*. It happened at $t \approx 10^{13}$ s, about 4×10^5 years.

The production of the first atoms was, of course, an important event in the evolution of the Universe. But this stage had an even greater significance for cosmology in that it was associated with a change in the transparency of the Universe. Before this, the electrons had been chiefly responsible for scattering the radiation. Since neutral gases are generally transparent, like the air around us, the formation of atoms corresponded to a sudden clearing of the Universe. For the first time one could have seen clearly through it — except of course that there was no life of any sort to see what was happening! Not only was it still too hot for living things to exist, but there were still no stars, nor were there any planets for them to have lived on. No structure yet existed because any pre-existing irregularities in the density of nucleons had been prevented from growing by the photons, which acted constantly to smooth the density out. There were pre-existing density irregularities in the dark matter, but these needed time to translate themselves into irregularities in the density of baryonic matter, which led to the structures we see today.

The temperature at decoupling was about 3000 K. The microwave background temperature today is about 2.7 K. Therefore the photon gas has redshifted by a factor of about 1100 since decoupling. This is then the factor by which the Universe has expanded since then. The density of matter in the Universe has changed by the inverse cube of this factor, about 10^{-9}. If we take the mass density today to be that required for $k = 0$ (10^{-26} kg m^{-3}) then this density at decoupling was about 10^{-17} kg m^{-3}. This is already a very low density compared to everyday densities on the Earth, which are 10^{20} times larger. Forming the Earth has therefore required a very great concentration of matter at later times!

7.6   The era of galaxies: from 10 million years to the present

The times given for the beginning of each era are all somewhat arbitrary because in reality the various processes merged into each other. The beginning of the galactic era is particularly uncertain. A galaxy is formed by one or more clouds of gas pulling themselves together by gravity and, since gravity is acting all the time, this gravitational condensation must be a continuous process; there is no sudden start to the growth of galaxies. However, the problem in setting the age of the galaxies is more serious than this semantic one. The manner and rate at which galaxies grow out of any small inhomogeneities in the gas are poorly understood. It is interesting that one of the most uncertain stages in the history of the Universe is the most recent one, and the one which only needs classical mechanics to reconstruct the processes. The observational way to settle the question would be to find half-formed galaxies. Presumably if we could look far enough away, and therefore far enough back in time, such objects must be visible. But to date, despite several tantalizing results, such searches have been unable to reach back far enough in time to clearly reveal the early formation of a galaxy.

Since the theoretical and experimental constraints are so weak, little more can be said than that it is thought that noticeable structure in the Universe probably *began* somewhere between 10^7 and 10^9 years after the big bang. Galaxies as we see them today can hardly have formed in less than 10^9 years. There is even debate over whether stars sometimes formed *before* galaxies, the stars only later gathering together to form galaxies. However, since a few stars can be seen in the very process of formation in our Galaxy at the present time, it is generally assumed that most stars are formed within existing galaxies.

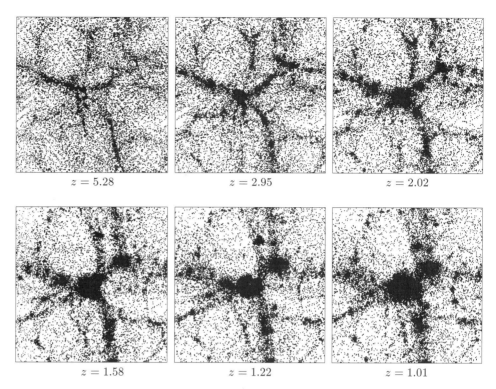

$z = 5.28$ $z = 2.95$ $z = 2.02$

$z = 1.58$ $z = 1.22$ $z = 1.01$

Figure 17   A computer simulation showing how initial slight inhomogeneities in the density distribution can lead to more pronounced inhomogeneities as matter is accreted from adjacent regions. Each square represents a two-dimensional projection of a comoving cube.

An understanding of galaxies and the development of clusters of galaxies can only be arrived at by carrying out detailed calculations. For instance, Figure 17 shows a computer simulation of the growth of pronounced inhomogeneities from what start out as only slight variations in density. This is done through the relatively dense regions attracting more and more matter from neighbouring areas. The sequence commences with a redshift $z = 5.28$ (less than 10^9 years after the big bang) and ends at a redshift $z = 1.01$ (about 3×10^9 years later). Its final shape is fairly typical of that of many clusters and superclusters.

Cosmologists would like to find out more about these processes because it would help to answer several questions:

- What did galaxies grow from and what were the initial inhomogeneities like?

- Is there a recognizable sequence in the evolution of galaxies, analogous to the well known sequence in the evolution of stars? (If so, one might be able to estimate the luminosity and hence distance of a galaxy from its type, by analogy with the way that the distances of stars can be judged by knowing their type and flux density — e.g. Cepheid variables, novae, supernovae. Such support for the 'distance ladder' is badly needed.)

- How does the luminosity of a galaxy vary with its age? (As was mentioned in Unit 13, uncertainty on this point seriously hampers interpretation of the Hubble plot.)

- What sorts of bodies are produced in galaxies — we can see the stars, but how much else is there that we cannot see?

It is the last two questions which lead us back to a reconsideration of the inventory of the contents in the Universe.

7.7   From the present into the future and domination by dark energy

The future of the Universe depends on its energy density, and on its contents. For instance, if the density has the critical value, but there is no dark-energy contribution, then the expansion of the Universe will continue forever, but at a gradually slowing rate. Eventually the rate of expansion will approach zero, but only in the *very* distant future.

Before that, in about 10^{12} years from now, all the light elements in the interstellar medium will have been used up in stellar nuclear reactions, and from then on, no new stars will form. As a consequence, all life throughout the Universe will die. The remnants of burnt-out stars and of planets will suffer collisions that take them towards the centre of their galaxy where they will be swallowed up by the black hole presumed to exist there. Even if they do not suffer such collisions, as they orbit the galactic centre, they will continuously emit gravitational radiation. This will result in them slowly spiralling in towards the centre and being captured by the black hole in that way. Here we are talking of a time-scale of about 10^{27} years.

Additionally we have to remember that the galaxies are themselves moving about as members of clusters, so they too will be losing energy via gravitational radiation. This could mean that each cluster will end up as a supergalactic black hole in some 10^{31} years.

So, does that mean the ultimate fate of a matter-dominated universe is that it will consist of a collection of black holes? Not necessarily. We have seen how Stephen Hawking has pointed out an interesting quantum-physical mechanism by which black holes can themselves radiate energy. Given sufficient time, a black hole 'evaporates'. Black holes left by burnt-out stars are expected to take 10^{67} years, the more massive ones at galactic centres 10^{97} years, and supergalactic ones 10^{106} years. At the end of that time the Universe will consist of nothing but cold, dilute radiation. This will mark the so-called *heat death of the Universe.*

heat death of the Universe

Finally we remark on the fate of a universe dominated by dark energy. As for an Einstein–de Sitter universe, or a matter-dominated universe with negative spatial curvature, superclusters of galaxies will become isolated by cosmic expansion, if dark-energy dominates. However, the rate at which this will happen is radically increased by dark energy. If the Universe is spatially flat and dominated by matter, then distances between superclusters will increase by a factor of order $100^{2/3} \approx 22$ in the next 1.4×10^{12} years, provided that matter-domination persists. If the formula for H^2 is dominated by negative spatial curvature, then the corresponding expansion factor will be of order 100. If, as is currently suggested, dark energy is already dominant, then the expansion factor will be more like the monstrous looking $\exp(100) \approx 10^{43}$. Dark energy, while *locally* undetectable, would thus lead to superclusters becoming cosmically isolated far sooner than was imagined by twentieth-century cosmologists.

Fortunately, humankind will not need to survive for anything like 10^{12} years to settle the issue of the eventual rate of expansion of the Universe. This is likely to be illuminated by refinements of measurements over a few decades. By 2015, the centennial year of Einstein's field equations of general relativity, one might hope to have a rather good idea of whether the constant term that Einstein introduced into cosmology – and then discarded – does indeed determine the fate of the Universe. The clues are probably to be found in signals from the most distant objects that we can study, at redshifts of order $z = 10$, in the case of the first protogalaxies, or of order $z = 1000$, in the case of the last interactions of the cosmic microwave background radiation with matter. Perhaps the study of the interactions of particles at the highest energies achievable on planet Earth will also contribute to this debate, by revealing the nature of dark matter and by suggesting a mechanism by which Einstein's cosmological constant might have been generated by changes in the structure of the vacuum, as has been speculated by particle physicists. However it turns out, it will be a privilege to live through the next few years of our attempt to understand the violent beginnings, sedate middle age and presumed cold fate of the Universe as a whole.

8 Epilogue

The picture of cosmology we have presented in this Block is remarkably detailed, even audacious in places. Not only do cosmologists have explanations for where the helium and other light elements came from, but they seriously expect to be able to explain *why* the Universe is so homogeneous and isotropic. In doing so, they are joining forces with high-energy particle physicists. The common ground between the two fields is considerable. The early Universe is the only place where the highest-energy interactions have ever taken place. The success of big bang theory in predicting the number of kinds of neutrinos did a lot for the confidence that physicists have in using cosmology as a testing ground for new physical theories.

Recent developments that have produced an increasing degree of agreement between the very precise results obtained from cosmic background anisotropies, the observation of Type Ia supernovae and the studies of the evolution of structure, have greatly boosted the confidence of cosmologists. But many challenges remain and there can be no doubt of the need for further work in all aspects of cosmology.

Physicists and astronomers will continue trying to discover the nature and quantity of dark matter. They will also continue to investigate dark energy speculating about its nature while testing its existence. All of this will get us closer to discovering whether inflation really happened, whether cosmic strings are really out there, even what really happened when the Universe was formed. Fundamental questions like these would once have been regarded as unanswerable. But that is no longer the case. The main lesson of the development of cosmology since the discovery of the cosmic microwave background is that it has proved possible to *understand* the Universe; we no longer just observe it.

Summary

1 The agents that cause the curvature of spacetime are (i) the energy density, ρ; (ii) the momentum per unit volume; and (iii) the momentum flux.

2 Because of the homogeneity of space, the value of the first of these contributions ρ, is the same throughout space.

3 Because of the isotropy and homogeneity of the Universe, the second of these contributions (momentum density) is zero.

4 Because of the uniform expansion of the Universe, the third contribution (momentum flux) is described by a single quantity p, referred to as the pressure.

5 Applying Einstein's field equations to a universe described by the Robertson–Walker metric yields the Friedmann equations (Equations 1 and 2). These link the quantities R and k of the metric to the two agents producing curvature, ρ and p.

6 The energy density, ρ, is made up of a contribution due to radiation, $\rho_{\mathrm{radiation}}$, one due to matter, ρ_{matter}, and one due to dark energy, ρ_Λ. Because the first two of these contributions vary respectively as $1/R^4$ and $1/R^3$ and the third contribution is constant, the Universe has gone from a period of radiation-dominance in the early stages, to matter-dominance at a later time and is now thought to be in a period of dark-energy-dominance.

7 It is clear from the Friedmann equations that $\mathrm{d}^2R/\mathrm{d}t^2$ must be negative during eras of radiation-dominance and matter-dominance, but it should be positive during a period of dark-energy-dominance.

8 $\mathrm{d}R/\mathrm{d}t$ could be of either sign, but Hubble's results show that it is currently positive. This gives rise to several possibilities:

(i) $\mathrm{d}R/\mathrm{d}t$ continues to be positive, corresponding to a universe whose curvature continues to get weaker and weaker, indefinitely. This possibility is known as an 'open universe' — a universe that will increase in size forever. In such a case, space has a negative curvature and is infinite in extent.

(ii) $\mathrm{d}R/\mathrm{d}t$ continues to be positive, but asymptotically approaches zero in the infinite future. This corresponds to a spatially flat universe with no dark energy.

(iii) $\mathrm{d}R/\mathrm{d}t$ changes sign and then becomes increasingly negative, so that R will eventually drop to zero. This possibility is known as a 'closed universe' — a universe that will reach a maximum size and then start to contract again. In such a case, space has a positive curvature and may 'close round' on itself as on the surface of an expanding balloon.

(iv) $\mathrm{d}R/\mathrm{d}t$ continues to be positive but grows bigger at an ever increasing rate. This possibility is known as an 'accelerating universe' and could be explained by the dominance of dark energy.

9 The inverse of the Hubble parameter H_0, provides an upper limit to the age of a matter-dominated universe, currently estimated to be 1.4×10^{10} years. This is close to the actual age of our Universe, which is thought to be dominated by dark energy.

10 In a matter-dominated universe with flat spatial geometry, we get the standard Einstein–de Sitter model with

$$R(t) \propto t^{2/3}$$

and $t_0 = \dfrac{2}{3}\dfrac{1}{H_0}$.

11 A number of different lines of evidence, including studies of the cosmic microwave background radiation, observations of Type Ia supernovae and attempts to model the observed large-scale clustering of galaxies, all indicate that our Universe contains a significant amount of dark energy in addition to matter and radiation.

12 According to the WMAP results (2003) the Universe has a total density that is consistent with the current critical value; about 73% of that density is due to dark energy and **27**% is due to matter. Only about one-sixth of the matter (about 4 or 5% of the total density) is thought to be due to baryonic matter.

13 Astronomical studies of the nature and distribution of matter also indicate that there is insufficient matter in the Universe to entirely account for the critical density. Such studies are difficult to perform and suffer from large uncertainties, but they also confirm that much of the matter that is present cannot be baryonic. The nature of this non-baryonic dark matter is an important issue.

14 Dark matter might consist of weakly interacting massive particles (WIMPs). Massive compact halo objects (MACHOs) and massive neutrinos are also expected to contribute, but their contributions are believed to be small.

15 The suggestion that the Universe underwent a brief period of super-fast expansion called inflation, between 10^{-34} s and 10^{-32} s after the big bang would solve three important problems: (i) the density is close to critical; (ii) the Universe is more or less homogeneous and isotropic on the large scale and (iii) there are no magnetic monopoles.

16 It is believed that there was no time before the big bang, because both time and space were formed at the instant of the big bang.

17 In Section 7, a brief history of the evolution of the Universe is given, highlighting the various distinctive phases through which the Universe is believed to have passed. An indication is also given of the possible future development of the Universe.

Band 6 of AC4 comments on this Unit.

Acknowledgements

Grateful acknowledgement is made to the following sources for permission to reproduce material used in this unit:

Figure 10 courtesy of Richard Sword, Cambridge University. Figure 11 courtesy of Royal Observatory, Edinburgh. Figure 12: Jeff Hester and Paul Scowen (Arizona State University) and NASA. Figure 13 REPRINTED WITH PERMISSION FROM SCHWARZSCHILD, B. 'MACHO SEARCHES FIND MOST CANDIDATES IN UNEXPECTED PLACES', *PHYSICS TODAY*, JULY 1994, P. 18, COPYRIGHT 1994 AMERICAN INSTITUTE OF PHYSICS. Figure 14 courtesy of William Forman and Christine Jones, Harvard–Smithsonian Center for Astrophysics, Cambridge, Mass. Figure 17 credit Frank J. Summers, Princeton University Observatory. Photo of Einstein and Hubble courtesy of The Huntington Library, San Marino, California.

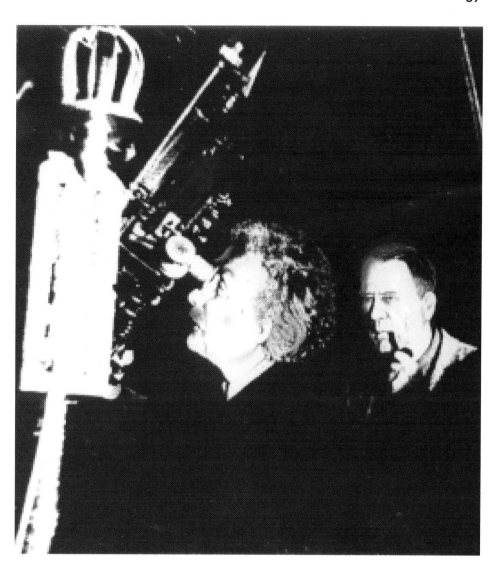

Self-assessment questions — answers and comments

SAQ 1 In a homogeneous universe, Einstein's field equations predict the geometry of spacetime in terms of two numbers: the density of energy ρ, and the pressure p. In our Universe we are assuming that the major contribution to ρ and p is associated with the mass and momentum flux carried by hydrogen atoms. But if it were possible to construct another universe with the same values of ρ and p, the geometry of spacetime would remain unchanged. This is because Einstein's field equations contain only quantities like energy and momentum — the specific *agencies* carrying these quantities, whether matter, antimatter or radiation, are irrelevant.

Moreover, the principle of the universality of free fall tells us that, once the geometry of spacetime is fixed, all test particles will follow similar trajectories, regardless of their composition. We conclude that the motions of clusters of galaxies in a gold or antigold or antihydrogen universe would be exactly the same as in our own Universe, provided the values of ρ and p were unchanged.

SAQ 2 Equations 1 and 2 become respectively

$$\frac{1}{R}\frac{d^2R}{dt^2} = -\frac{4\pi G}{3c^2}(\rho + 3p)$$

$$\frac{1}{R}\frac{d^2R}{dt^2} + \left(\frac{1}{R}\frac{dR}{dt}\right)^2 + \frac{kc^2}{R^2} = \frac{4\pi G}{3c^2}(\rho - 3p).$$

SAQ 3 No. For any particular universe, k retains whatever value it has: $+1$, -1, or 0.

SAQ 4 Equations 1 and 2 can be written in the form:

$$-\frac{4\pi Gp}{c^2} = \frac{1}{R}\frac{d^2R}{dt^2} + \frac{4\pi G\rho}{3c^2}$$

$$-\frac{4\pi Gp}{c^2} - \left(\frac{1}{R}\frac{dR}{dt}\right)^2 - \frac{kc^2}{R^2} = \frac{1}{R}\frac{d^2R}{dt^2} - \frac{4\pi G\rho}{3c^2}.$$

The pressure p, can be eliminated by subtracting the second equation from the first. We find

$$\left(\frac{1}{R}\frac{dR}{dt}\right)^2 + \frac{kc^2}{R^2} = \frac{8\pi G\rho}{3c^2}.$$

Multiplying by R^2 we finally obtain

$$\left(\frac{dR}{dt}\right)^2 + kc^2 = \frac{8\pi G}{3c^2}\rho R^2$$

which is Equation 5 in the text.

Since $H = \frac{1}{R}\frac{dR}{dt}$, H can be zero if $\frac{dR}{dt}$ is zero, that is if

$$k = \frac{8\pi G R^2 \rho}{3c^4}.$$

Since G, R^2, c^4 and ρ are necessarily positive, this can only be satisfied if k is $+1$.

SAQ 5 Under the conditions that p is negligible. Equation 1 becomes:

$$\frac{1}{R}\frac{d^2R}{dt^2} = -\frac{4\pi G\rho}{3c^2}.$$

Since R, G and ρ are by their nature positive, d^2R/dt^2 must be negative, corresponding to a decreasing value for dR/dt, i.e. a decreasing rate of expansion.

SAQ 6 When $k = 0$, Equation 5 gives

$$\frac{1}{R^2}\left(\frac{dR}{dt}\right)^2 = \frac{8\pi G\rho}{3c^2}.$$

But the current value of the Hubble parameter is defined by $H_0 = \left(\frac{1}{R}\frac{dR}{dt}\right)_{t=t_0}.$

Substituting, we get

$$H_0^2 = \frac{8\pi G\rho_0}{3c^2}.$$

Thus

$$\rho_0 = \frac{3c^2 H_0^2}{8\pi G} \quad \text{if } k = 0.$$

SAQ 7 Yes — the same argument applies as in SAQ 5. dR/dt is negative in the contracting phase and has an increasingly large negative value, corresponding to negative d^2R/dt^2.

SAQ 8 (a) Variation of ρ for $k = +1$.

In a closed universe R rises to a maximum and then decreases. Thus ρ starts from a high value, decreases to a minimum (which coincides with a maximum value for R), and then returns to a high value.

As described in the text, the behaviour of the Universe in the first few seconds after the big bang can only be conjectured, so an initial small gap is left in the graph (Figure 18). The behaviour in the final stages of recompression is also conjectural, so a second gap is left before the dashed portion which indicates a possible re-expansion.

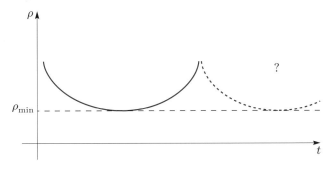

Figure 18 Conjectured variation of density with time, for a closed universe ($k = +1$).

The lowest value of ρ occurs when $\dfrac{\mathrm{d}R}{\mathrm{d}t} = 0$. Since

$$\left(\frac{\mathrm{d}R}{\mathrm{d}t}\right)^2 + kc^2 = \frac{8\pi GR^2\rho}{3c^2} \qquad (5)$$

it follows that

$$\rho_{\min} = \frac{3kc^4}{8\pi GR_{\max}^2}$$

(b) Variation of ρ for an open universe with $k = -1$.

In this case the value of ρ *always* decreases with time. There is no maximum value of R and the minimum value of ρ is zero, reached asymptotically (Figure 19).

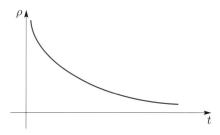

Figure 19 Conjectured variation of density with time, for an open universe ($k = -1$).

SAQ 9 The differences are not expected to be very dramatic. Our galaxy will be little affected, since it is dominated by its own gravity rather than the overall curvature. Looking at other galaxies with a telescope, the main change will be that spectral lines will be blueshifted rather than redshifted, for those galaxies whose observed light was emitted after the peak. Looking further into space the observer will come to galaxies whose light was emitted before the peak, and these will still be redshifted. There will, of course, be a spherical shell of galaxies with no shift at all.

The approaching galaxies will not only be blueshifted but will also be brighter than they are now when they are moving away, because a given energy output will be compressed into a shorter time rather than pulled out into a longer time. In other words the rate of receiving photons will increase with time.

SAQ 10 $\quad t_0 < \dfrac{1}{H_0} = \dfrac{1}{2.3 \times 10^{-18}\,\mathrm{s}^{-1}}$

So

$$t_0 < 4.3 \times 10^{17}\,\mathrm{s}$$

i.e.

$$t_0 < 1.4 \times 10^{10}\,\text{years}.$$

SAQ 11 Known upper and lower limits for the age of the Universe are shown in Figure 20.

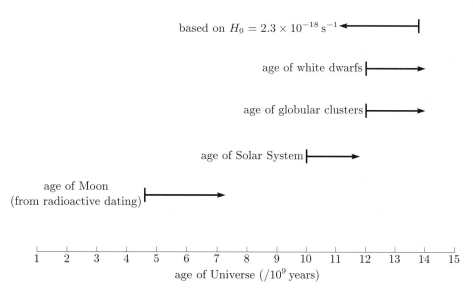

Figure 20 Known upper and lower limits for the age of the Universe.

SAQ 12 (a) A matter-dominated universe has

$$\rho \propto \frac{1}{R^3}$$

and a spatially flat universe has $k = 0$. Hence Equation 5 becomes

$$\left(\frac{dR}{dt}\right)^2 = \frac{8\pi G}{3c^2}\rho R^2 \propto \frac{1}{R^3}R^2 = \frac{1}{R}$$

or $R\left(\dfrac{dR}{dt}\right)^2 = $ a constant.

(b) If $R = At^{2/3}$,

$$R\left(\frac{dR}{dt}\right)^2 = At^{2/3}(A\tfrac{2}{3}t^{-1/3})^2$$
$$= \tfrac{4}{9}A^3$$
$$= \text{a constant.}$$

Therefore $R = At^{2/3}$ is a solution to the differential equation.

(c) The Hubble parameter is given by

$$H = \frac{1}{R}\left(\frac{dR}{dt}\right) = \frac{1}{At^{2/3}}(A\tfrac{2}{3}t^{-1/3}) = \frac{2}{3t}.$$

Hence $\qquad t = \dfrac{2}{3H}$

and in particular

$$t_0 = \frac{2}{3H_0}. \qquad (10)$$

(d) The current estimate of H_0, $2.3 \times 10^{-18}\,\text{s}^{-1}$, yields (through its reciprocal) an upper age limit to the Universe of 1.4×10^{10} years. However, Equation 10 shows that for a universe of the type being considered, its actual age is given by $2/3$ times the value of this upper limit, i.e. 0.95×10^{10} years.

SAQ 13 From SAQ 12(b) we have $R = At^{2/3}$. By differentiation,

$$\frac{dR}{dt} = \frac{2}{3}At^{-1/3}.$$

Differentiating a second time:

$$\frac{d^2R}{dt^2} = -\frac{2}{9}At^{-4/3}.$$

Substituting for R, (dR/dt), and (d^2R/dt^2) in Equation 14 of Unit 14 yields $q = 1/2$.

SAQ 14 From Equation 11 of Unit 14, we have

$$z = \frac{R(t_0)}{R(t_1)} - 1.$$

Using Equation 9 of this Unit, together with the fact that $z \gg 1$, we have, for the case where $t_1 = t_{\text{dec}}$ (decoupling time),

$$1000^{3/2} = 10^{10}/t_{\text{dec}}.$$

Thus, $t_{\text{dec}} \approx 3 \times 10^5$ years.

SAQ 15 In the case of the Sun, Equation 11 is

$$M_\odot = \frac{v^2 r}{G}.$$

The Earth takes a year $(3.156 \times 10^7\,\text{s})$ to travel a distance $2\pi \times 1\,\text{AU} = 9.4 \times 10^{11}\,\text{m}$. Thus its orbital speed is $v = 2.979 \times 10^4\,\text{m s}^{-1}$. Putting this and the value for r into the above equation yields the desired result.

SAQ 16 Allowing for the mass-to-light ratio of 10, we have found the mass density due to the luminous regions of galaxies to be $2 \times 10^{-28}\,\text{kg m}^{-3}$. Dividing by the visible mass of a typical galaxy, $2 \times 10^{41}\,\text{kg}$, we get a density of 10^{-69} galaxies per cubic metre. The average volume occupied by a single galaxy is the reciprocal of this, $10^{69}\,\text{m}^3$ per galaxy. The radius r of an equivalent sphere is found by setting this equal to $4\pi r^3/3$, giving $r = 6 \times 10^{22}\,\text{m}$. The mean distance between two galaxies would therefore be twice this, or $12 \times 10^{22}\,\text{m}$. This is 400 times the given typical size of a galaxy. The separation between our Galaxy and Andromeda is small, about one-eighth of the normal separation. This is part of the reason why astronomers believe that the two galaxies form a gravitationally bound pair.

SAQ 17 (a) $\quad \dfrac{kc^2}{R^2} = \dfrac{8\pi\rho G}{3c^2} - H^2. \qquad (6)$

We take

$$\frac{\rho_0}{c^2} = 1.0 \times 10^{-27}\,\text{kg m}^{-3}$$
$$H_0 = 2.3 \times 10^{-18}\,\text{s}^{-1}.$$

Then

$$\frac{kc^2}{R^2} = \left[\frac{8\pi}{3} \times 10^{-27} \times 6.67 \times 10^{-11} - (2.3 \times 10^{-18})^2\right]\text{s}^{-2}$$
$$= (0.56 \times 10^{-36} - 5.3 \times 10^{-36})\,\text{s}^{-2}$$
$$= -4.7 \times 10^{-36}\,\text{s}^{-2}.$$

Thus

$$\frac{k}{R^2} = -5.2 \times 10^{-53}\,\text{m}^{-2}.$$

The negative sign corresponds to negative curvature. (The value happens to be of the same order of magnitude as the inverse square of the distance to the furthest observed galaxies. But that is a coincidence, k/R^2 could have any value — including zero.)

(b) The mass density which makes $k = 0$ is

$$\frac{\rho_0^c}{c^2} = \frac{3H_0^2}{8\pi G} = \frac{3(2.3 \times 10^{-18})^2}{8\pi(6.67 \times 10^{-11})} = 9.5 \times 10^{-27}\,\text{kg m}^{-3}.$$

SAQ 18 Zero curvature $(k = 0)$ requires a present mass density of $9.5 \times 10^{-27}\,\text{kg m}^{-3}$ (see SAQ 17 part b). If this was the present baryonic density it would correspond to a freeze-out mass fraction of deuterium that is less than 10^{-11} (see Figure 29 of Unit 13); *much* less than the estimated value of $(3\,\text{to}\,7) \times 10^{-5}$.

SAQ 19 The amount of helium produced has only a weak dependence on the density, so is not a sensitive way of finding the density. (This argument might lose its force if the abundance of helium happened to be much easier to measure than the abundance of deuterium. However, this is not the case. Both helium and deuterium are rather hard to identify, and it is difficult to take a suitably weighted average over very large volumes for either of them.)

SAQ 20 Before the reaction, the combined mass of the proton and neutron was 3.3473×10^{-27} kg. The difference between this and the mass of the final deuteron is $(3.3473 - 3.3433) \times 10^{-27}$ kg $= 0.0040 \times 10^{-27}$ kg. As a fraction of the initial mass, this represents $0.0040/3.3473 \approx 1.2 \times 10^{-3}$. This is close to the figure of 0.1% quoted in the text.

SAQ 21 (a) We have two lengths that should be equal. The first is the radius of a black hole with energy, E. This is roughly GM/c^2 or GE/c^4. The second is the wavelength of a photon of the same energy, i.e. hc/E. (Since this is an argument designed to tell us what order of magnitude of distance we are talking about when we say a photon can spontaneously form a black hole, we ignore factors of 2, such as that which appears in the expression for the radius of the black hole, and 2π, the difference between h and $\hbar$.) Setting these distances equal gives

$$\frac{GE}{c^4} = \frac{\hbar c}{E}.$$

Thus

$$E_{\mathrm{Pl}}^2 = \frac{\hbar c^5}{G}.$$

This gives E_{Pl}, the Planck energy. The Planck mass, given in the text, is this energy divided by c^2. The Planck length scale is

$$r_{\mathrm{Pl}} = \frac{\hbar c}{E_{\mathrm{pl}}} \approx \left(\frac{G\hbar}{c^3}\right)^{1/2},$$

as given in the text. It is straightforward to put in values for the constants to find that $r_{\mathrm{Pl}} \approx 10^{-35}$ m.

(b) The time to travel a distance is the distance divided by the speed, so the Planck time is

$$t_{\mathrm{Pl}} = r_{\mathrm{Pl}}/c = \left(G\hbar/c^5\right)^{1/2} \approx 10^{-43} \text{ s.}$$

(c) The Planck mass is $m_{\mathrm{Pl}} = (c\hbar/G)^{1/2} = 2 \times 10^{-8}$ kg. The mass of a proton is $m_{\mathrm{p}} = 1.7 \times 10^{-27}$ kg. Their ratio is $m_{\mathrm{p}}/m_{\mathrm{Pl}} = 8.5 \times 10^{-20}$, so we have $m_{\mathrm{p}} \approx 10^{-19}m_{\mathrm{Pl}}$. The size of the proton is about $r_{\mathrm{p}} = 10^{-15}$ m, so we have $r_{\mathrm{p}} \approx 10^{20}r_{\mathrm{Pl}}$. The shortest decay lifetime of fundamental particles, $\tau_0 = 10^{-20}$ s, is given by $\tau_0 = 10^{23}t_{\mathrm{Pl}}$.

SAQ 22 (a) $\left(\dfrac{\mathrm{d}R}{\mathrm{d}t}\right)^2 = \dfrac{8\pi G}{3c^2}\rho R^2 - kc^2.$ \hfill (5)

For a matter-dominated universe, ρ depends on $1/R^3$. Thus the first term on the right-hand side approaches zero as R becomes large. With $k = -1$, the right-hand side simply becomes c^2, and this shows the rate of expansion ($\mathrm{d}R/\mathrm{d}t$) has become a constant.

(b) $H \equiv \dfrac{1}{R}\dfrac{\mathrm{d}R}{\mathrm{d}t}.$

We have already seen that at large R, ($\mathrm{d}R/\mathrm{d}t$) becomes a constant. Thus, $H \propto 1/R$ for large R.

(c) From Equation 7 we have

$$\rho^{\mathrm{c}} = \frac{3c^2}{8\pi G}H^2.$$

But we know from (b) that for large R, H varies as $1/R$. Therefore, $\rho^{\mathrm{c}} \propto 1/R^2$.

ρ on the other hand varies as $1/R^3$ due to the volume of the Universe increasing in response to the scaling factor R.

Thus, the ratio $\rho/\rho^{\mathrm{c}} \propto 1/R$.

SAQ 23 We shall show that the given solution, $R = A\mathrm{e}^{t/\tau}$, where A is a constant, satisfies Equation 12. Using the property of the exponential that

$$\frac{\mathrm{d}}{\mathrm{d}t}\mathrm{e}^{at} = a\mathrm{e}^{at}$$

for any constant a, we have on application of this rule twice

$$\frac{\mathrm{d}^2}{\mathrm{d}t^2}\mathrm{e}^{at} = a^2\mathrm{e}^{at}.$$

It follows, by setting $a = 1/\tau$, that

$$\frac{\mathrm{d}^2}{\mathrm{d}t^2}R = \frac{A}{\tau^2}\mathrm{e}^{t/\tau} = \frac{1}{\tau^2}R.$$

Therefore we have

$$\frac{1}{R}\frac{\mathrm{d}^2R}{\mathrm{d}t^2} = \frac{1}{\tau^2} = \frac{8\pi G\rho_{\mathrm{i}}}{3c^2},$$

which is Equation 12. So the expression given in Equation 13 is a solution of the equation governing the expansion of the Universe.

Note however that the demonstration given here depends in the end only on τ^2, not τ itself — it does not depend on the *sign* of τ. Therefore, there is another solution to Equation 12, yielding the general solution

$$R = A\mathrm{e}^{t/\tau} + B\mathrm{e}^{-t/\tau},$$

where A and B are constants. The values of A and B will be determined by the details of the starting conditions of the Universe at the time when this inflation begins, and we cannot know what these details are. But they don't matter to us, because after a significant amount of time, many times the characteristic time τ, the term multiplying B becomes much smaller than the term multiplying A. Unless by chance the starting conditions give A to be exactly zero, then after a time much longer than $t = \frac{1}{2}\tau \ln(B/A)$, the second term will be negligible and the Universe will follow the expansion law in Equation 13. The only exception is if $A = 0$ or if A is so small compared to B that the time $\frac{1}{2}\tau \ln(B/A)$ is longer than the time available for inflation. These are regarded as so unlikely for random starting conditions that the non-inflating solutions of Equation 12 do not need to be considered.

SAQ 24 The energy $E = 10^{14}\,\text{GeV}$ is equivalent to $1.6 \times 10^4\,\text{J}$. The characteristic length scale associated with photons of this energy is $\lambda = hc/E = 1.2 \times 10^{-29}\,\text{m}$. The energy density of vacuum fluctuations with wavelengths cut off at this energy rather than the Planck energy is therefore $E/\lambda^3 = 8.4 \times 10^{90}\,\text{J m}^{-3}$.

SAQ 25 The expression for τ is given by Equation 14:

$$\tau = \left(\frac{3c^2}{8\pi G \rho_v} \right)^{1/2}.$$

Substituting $10^{94}\,\text{J m}^{-3}$ for ρ_v yields

$$\tau \approx 1.3 \times 10^{-34}\,\text{s}.$$

SAQ 26 Once the balloon has been inflated, we find that an arbitrarily tiny area of its surface appears, to all intents and purposes, to be flat. The greater the degree of inflation, the closer the approximation to flatness — and this is so regardless of what the curvature of the balloon had been initially.

This can be used as an analogy for cosmological inflation if we equate the tiny area of the balloon's surface with the *observable* Universe. It too can be made as flat as one likes, regardless of its initial curvature before inflation. The implication is, of course, that the *observable* Universe is but a tiny part of the *overall* Universe. As time goes by, we are able to receive light from further away, and so the observable universe progressively encompasses an ever greater proportion of the whole — but it is still likely to remain only a tiny fragment of all that exists.

SAQ 27 When we speak of a flat geometry we mean that the curvature is essentially zero. For the Robertson–Walker metric, the expression for the curvature is k/R^2. This can be essentially zero *either* because $k = 0$, *or* because R^2 is very large. Through inflation R is made exceedingly large — to the extent that *regardless of the value k* the curvature will approach zero. So, the answer is no; inflation does not alter the value of k; it will only have the value zero if that is how it started out.

Unit 16 Consolidation and revision II

Prepared by the Course Team

Contents

Study Guide for Unit 16

This Unit generally follows the pattern of Unit 8, and is intended to help you refresh and deepen your understanding of the second half of the course. As in Unit 8, a number of consolidation questions are provided, and as in that Unit, they are more like TMA questions than exam questions in difficulty. That is to say, the exam will not contain questions which are as long as the longer consolidation questions, nor are we likely to set exam questions which require as much puzzling out as some TMA questions do. Examination questions are generally straightforward, and you should look at the Specimen Exam and the Solutions to the Specimen Exam for an indication of the structure of the exam and the level of difficulty of exam questions.

The exercises are not all of the same length and difficulty. While some are relatively short and straightforward like SAQs, others will require more thought, approaching the level of difficulty seen in TMA questions. The longer exercises are subdivided into short sections, a few of which are marked with an asterisk. *The asterisked sections are a little more demanding, and it is safe to ignore them, at least at the first reading.*

Use of asterisk * in questions.

Unit 16 does contain material of immediate relevance to the exam: notes for answers to essay questions.

Study comment

What to do if you are behind:

Jump straight to Sections 2 and 4 on comprehension exercises.

Then study Section 6 which is an overview of the course and could well be useful for answering essay questions.

Then study Section 7 on essay exercises.

Finally, browse through the appendices since they refer to some common stumbling blocks.

Even if you have no time to attempt many of the consolidation exercises, you will still find it useful to read them and their solutions.

---

1   Introduction

Unit 16 is designed to help you deepen and consolidate your understanding of the course. It contains no new concepts. Some parts, particularly the appendices, are designed to help you prepare for the examination.

Much of Unit 16 follows the pattern of Unit 8, but concentrating on Blocks 3 and 4. The comprehension exercises (whose answers are to be found in the Glossaries or in some cases, specified parts of Units) will check your awareness of definitions and basic concepts.

The consolidation exercises, as in Unit 8, reinforce understanding of certain key areas with particular examples and alternative approaches. They should be helpful for Part II of the examination, but should not be taken as indicating the style or standard of this part of the exam. Indeed some, like the harder parts of TMAs and CMAs, are intended to provoke your thoughts over a time-scale of days, not minutes. Do note that the consolidation exercises by no means completely cover all of the significant content of the Units.

In Part III of the examination, you will have a choice of essay questions which typically ask you to draw on your appreciation of a particular theme as it appears over the entire course. This is intended to test how well you have fitted the sections of the course together in your mind. (It is, of course, also designed to assess your ability to communicate through the medium of text, with equations and diagrams as needed. Don't forget that some advice on essay style and structure was included in Unit 8.) To help you form an integrated view of the course, a section is included which traces two themes from Block 1 through the three other Blocks. These are the twelve assumptions (A1–A12) of Newtonian space and time from Unit 1, and the principle of relativity from Unit 3. The essay exercises that follow require this integrative approach. The 'answers' are in the form of the type of skeleton structure you should sketch out before you start.

At this point, I would like to become a little more personal than is usual for the author of a Unit. The course team invited two experienced course tutors from S354, the precursor to S357, to join them to write Units 8 and 16. Naturally, as one of those invited, I consider this to have been an astute move. The final sections arise directly from my experience as a tutor. I have selected six topics from across the course, each of which I know presents real difficulties to a significant fraction of students, and in the first six appendices I have attempted to alert you to the problems and clarify the issues. It may be that none of these has ever troubled you or that you have rectified a misconception following a TMA return. But the chances are that two or more of these could still trip you up in the exam, so look at all the appendices and study them selectively. The final appendix suggests a revision strategy for the exam.

The final section of the Unit proper, entitled 'A tutor's perspective' is really a collage of the perspective of many S354 students, refracted through and interpreted by this particular tutor. The response of these students to that course has been overwhelmingly positive, and I have gained the strong impression that this is not primarily because they have learned a lot of physics, but because they have (often for the first time) had a much broader experience of what science is about and how it is done. I would like to convey something of this to the next generation of students, and if in doing so I mix in some personal views to illuminate this response, I hope these will be tolerated or even appreciated.

2 Comprehension exercises for Block 3

Complete answers to most of these exercises may be found in the Glossary for Block 3 by looking up the word or phrase displayed in capital letters. In other cases, Unit references to answers are given in square brackets.

1 In Blocks 1 and 2, an INERTIAL FRAME OF REFERENCE is defined as a frame of reference in which test bodies, not experiencing any force, move at constant velocity. In what crucial respect is this definition amended in Block 3?

2 FREE FALL is the motion of a ......-..... subject only to .... influences or to no influences at all.

3 What is meant by the UNIVERSALITY OF FREE FALL?

4 In Newtonian mechanics, the universality of free fall is explained by equating inertial mass to gravitational mass. In Block 3, it follows from the fact that GENERAL RELATIVITY is a .... theory of motion under gravity.

5 State two of the GEOMETRIC PROPERTIES of a two-dimensional space. How could you tell from these properties whether or not the space was Euclidean?

6 How can you determine whether you are in an ACCELERATING FRAME OF REFERENCE

(a) by examining clocks,

(b) by making geometric measurements?

7 Write down a statement of the EQUIVALENCE PRINCIPLE and its immediate consequences. Your response should contain the following phrases:

an accelerating frame of reference in the absence of any aggregate of matter;

a frame of reference at rest relative to a large aggregate of matter;

a freely falling frame of reference;

local region of spacetime;

precision of measurement.

8 In a LOCALLY INERTIAL FRAME OF REFERENCE, exactly *what* is restricted?

9 Write down the DISTANCE FUNCTION Δl in terms of

(a) changes in the Cartesian coordinates x^1 and x^2 of a (Euclidean) plane;

(b) changes in polar coordinates ρ and ϕ of a (Euclidean) plane;

(c) changes in generalized coordinates q^1 and q^2 of any surface. [For answer see Tables 4 and 5 of Unit 10 & 11.]

10 A curve in a two-dimensional space can be specified by two COORDINATE FUNCTIONS $q^1(s)$ and $q^2(s)$. What name is given to the variable s?

11 What is meant by the METRIC OF A TWO-DIMENSIONAL SPACE?

12 Write down the METRIC OF THE PLANE IN CARTESIAN COORDINATES.

13 Write down the METRIC OF THE PLANE IN PLANE-POLAR COORDINATES.

14 What properties of a space are determined by the METRIC?

15 What is a GEODESIC (a) in two-dimensional space, (b) in spacetime?

16 What are GEODESIC EQUATIONS? In which physical situations do these lead to (a) minimum, (b) maximum, intervals?

17 Write down the METRIC OF SPACETIME, for the spacetime of special relativity. What more general form applies in the presence of matter?

18 For neighbouring events, what does the interval given by the METRIC OF SPACETIME represent (a) for time-like separated events, (b) for space-like separated events?

19 In the language of the previous questions, how does the GENERAL THEORY OF RELATIVITY describe the motion of a test particle in free fall? What does the theory say in the case of a light pulse?

20 Briefly explain what is meant by the CURVATURE_i ($i = 1$ to 4) of a two-dimensional surface. Why is curvature_1 not illuminating with respect to general relativity? Why is curvature_4 more useful than curvature_3?

21 The CURVATURE_4 function, as obtained from the metric, has different forms depending on the choice of coordinate system. Does the *value* of the function, at any point in a two-dimensional space, depend on the choice of coordinate system?

22 Name two ways in which the GEOMETRIC PROPERTIES of a geometric figure drawn on a surface of non-zero curvature differ from the properties of the same figure drawn on a plane.

23 The spacetime of general relativity is characterized by two functions, the so-called RICCI CURVATURE and the RIEMANN CURVATURE.

(a) What are they functions of?

(b) How many independent component functions does each comprise?

(c) Which one is the analogue of curvature_4 for a two-dimensional space?

(d) If at a specified event only one of them is zero, which would it be?

(e) How can one be obtained from the other?

(f) Which is *immediately* determined by the source terms in Einstein's field equations?

24 Name the three types of contribution to the SOURCE TERMS of Einstein's field equations.

25 Explain the significance of the SCHWARZSCHILD SOLUTION of Einstein's field equations.

26 Explain the term 'loss of METRICAL SIGNIFICANCE' making reference to the coordinates (t, r, θ, ϕ) in which the SCHWARZSCHILD metric is generally written.

27 What name is given to the radius of the EVENT HORIZON surrounding a non-rotating black hole?

28 What happens to the light from an object as it approaches the EVENT HORIZON?

29 What is the special characteristic of the PHOTON-SPHERE of a non-rotating black hole?

30 What is the name of the entity at the centre of a spherical black hole? Why is it not possible for an object which has crossed the event horizon to crash into the surface of a large sphere? [See Section 2.7 of Unit 12.]

31 Would a freely falling observer notice anything odd on passing through the event horizon of a black hole? What physical effect would a real observer notice on falling into (a) a stellar black hole, (b) a galactic-centre black hole, and where (in relation to the event horizon) would it be noticed? [See Section 2.6 of Unit 12.]

32 Explain

(a) why 'mini black holes' emit significant Hawking radiation while stellar or galactic-centre black holes do not,

(b) why such mini black holes will eventually destroy themselves explosively. [See Section 4 of Unit 12.]

33 What is the distinctive feature of the motion of an object within the ergosphere of a Kerr black hole? In what direction should you approach such a black hole in order to get as close as possible, assuming you wish to escape along the same path as you approached? [See Section 3 of Unit 12.]

34 As a consequence of Birkhoff's theorem, what property must a supernova have in order to emit gravitational waves? [See p. 42 of Unit 12.]

3 Consolidation exercises for Block 3

In the consolidation questions and in the corresponding answers, all references by number to equations, figures and SAQs refer to the equations, etc. in the corresponding Unit.

3.1 Unit 9

In Blocks 1 and 2, we assumed that global inertial frames existed, and took Euclidean geometry for granted. We also assumed that, in a given inertial frame, clocks everywhere run at the same rate. The laws of physics are formulated for inertial frames. We know that we cannot rely on them in non-inertial frames.

Unit 9 starts by looking at the properties of space and time as seen from a rotating frame and asserts that gravity induces similar changes. Section 3.1 considers a rotating space station and shows that only in an extended laboratory within the station can the effects of the acceleration be distinguished from the gravitational effects of a possible distribution of matter. You will have noticed that the force that apparently acts on stationary objects in the station is directed outwards from the rotation axis. Of course a spherically or axially symmetric distribution of matter generates an inwards gravitational force. This raises the question of whether the two effects can be cancelled.

Exercise 1 Suppose we assemble the space station from a very large number of smaller sections each in a circular orbit of the same radius around a planet. Is it possible for an experimenter in this laboratory to deduce that it is indeed rotating around a massive object, rather than travelling at uniform velocity without rotation in outer space?

[Hint: can 'gravity' be exactly cancelled everywhere?]

Exercise 2 Which one of the following can be said to be in free fall, to a good approximation? Why do the others not qualify?

(a) A parachutist.

(b) Either of a binary star pair consisting of two stars of comparable mass, orbiting their mutual centre of mass.

(c) The planet Mercury.

Exercise 3 The experimenter in the freely falling elevator (Figure 14 of Unit 9) releases two particles from rest but in this case one is above the other. Given sufficient spacetime and instrumental accuracy, what will be observed?

In Section 4, we arrive at Einstein's insight into the significance of the universality of free fall (or, if you like, the equality of gravitational and inertial mass). The effects of gravity and of an accelerating reference frame are indistinguishable because they are the same, inside a sufficiently small spacetime volume. The (weak) equivalence principle, superficially a more-or-less self-evident statement of experience of simple physics, becomes a window into the new physics and allows us to predict effects such as the gravitational redshift.

Exercise 4 A space laboratory is accelerating far from any aggregates of matter. A laser in the laboratory fires a beam in the direction of this acceleration. How does the light received at the other end of the laboratory differ from the emitted light? What does the equivalence principle tell you from this?

Exercise 5   Referring to SAQ 16, derive an expression for the deflection P_1P_2 of the laser beam at the Earth's surface (under the conditions specified in Part (b)) in terms of the width w of the laboratory and the acceleration due to gravity at the Earth's surface. To what accuracy could a 100 m horizontal line on the Earth possibly be described as 'straight'?

[Hint: make use of Equation 7 of Unit 9.]

3.2 *Unit 10 & 11*

Unit 10 & 11 is concerned both with general relativity itself and with the two-dimensional analogy or allegory. The allegory is given a fairly full mathematical treatment, while the corresponding mathematics of general relativity are only alluded to or presented in outline. We have seen how very careful we must be when translating quantities like curvature between the allegory and general relativity. Appendix 6 contains further advice in this area. The Unit tends to refer back and forth rather frequently so that the relevance of the allegory is not lost; here we shall concentrate on each separately and perhaps gain some clarity.

Although the allegory is conceptually simpler than general relativity, you may well find that mastering it presents you with greater difficulties. This is almost certain to be the case if you are having problems following the mathematics. Many of the equations seem long and intractable so that you might have little confidence in applying them. If this applies to you then you may find the notes in Appendix 5 (Mathematical hazards) helpful. Be assured that you will *not* need to memorize such intricate equations as the geodesic equations (38a, b, c) or the expression for curvature$_4$. But such equations might be presented in an exam, and the following exercises should help give you appropriate practice in using them.

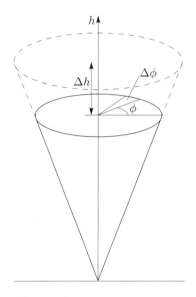

Figure 1   Inverted right circular cone.

Exercise 6   The metric of the surface of an inverted right circular cone (see Figure 1) is given by

$$\Delta l = [(1 + k^2)(\Delta h)^2 + (hk\,\Delta\phi)^2]^{1/2} \tag{1}$$

where h is the height above the apex, ϕ is the angle measured around the axis, and k is constant.

(a) Refer to Equation 33 and identify the coordinates q^1 and q^2 and the metric coefficients g_1 and g_2.

(b) Determine the curvature function $\mathscr{K}$ using Equation 68b. Is your answer reasonable?

A line is defined by

$$h = a/((1 + k^2)^{1/2}\cos s), \qquad \phi = s(1 + k^2)^{1/2}/k \tag{2}$$

a being constant.

(c) Use Equation 38c to show that $L = a/\cos^2 s$.

(d) Prove that the line satisfies the geodesic equations (38a) and (38b).

[Remember: you will not be asked to recall the geodesic or curvature equations in the exam, but such equations might be presented to you, in which case you will be expected to interpret and manipulate them.]

That exercise covers a large section of Unit 10 & 11 material: Sections 3, 4, 6, 7 and 12 are all relevant to it. However, before we leave the two-dimensional allegory, we should examine the relationship between curvature and geometric properties explored in Section 5 and the end of Section 6. Much of the discussion concerns figures drawn on a spherical surface. Because such a surface has constant curvature, the treatment can be quite simple.

Exercise 7   In SAQ 10, you found that the ratio of circumference to radius for a circle of maximum circumference drawn on a sphere is 4. What happens if the radius of the circle increases further?

Exercise 8   Derive an expression for the ratio of the area enclosed by a circle drawn on a sphere to the square of its radius measured in the surface of the sphere. What is this ratio for a very small circle?

[Hint: $\cos\theta \approx 1 - \theta^2/2$ for small θ.]

*   **Exercise 9**   If a triangle drawn on the surface of a sphere is made larger and larger, then in the limit, it will cover the whole surface area of the sphere. What is the sum of the *interior* angles of such a triangle?

Exercise 10   All lines of latitude cut the meridians at $\pi/2$. Hence a 'rectangle' formed between any two latitude lines and any two meridians will have angles summing to 2π, just as on a plane. What is the flaw in this argument?

3.3   Unit 12

At the end of Unit 10 & 11 we met the Schwarzschild metric — the solution to Einstein's field equations for the spacetime outside an isolated spherically symmetric mass distribution of mass M. We saw that it predicts the same radial motion for freely falling particles as does Newton's law of gravitation, in the limit of the radial distance being much greater than GM/c^2 and the speed much less than c. However, it does predict slight departures from Newtonian expectations such as the bending of light around stars and the delay in echoes from planets. In Unit 12, we remove the restriction $r \gg GM/c^2$ and confront not only much more extreme departures from rectilinear light propagation but also very strange particle dynamics. In particular, if the mass M is contained within the surface (the event horizon) of radius $2MG/c^2 \equiv k$ then we have a black hole. No matter or radiation, however energetic, can escape from within the event horizon of a black hole.

Exercise 11   Would it be possible to build a tower on a black hole extending above the event horizon, perhaps from in-falling matter?

Even outside the event horizon, both space and spacetime are highly curved, so neither Euclidean geometry nor Newtonian or special relativistic spacetime properties are at all accurate (for large spacetime volumes) until $r \gg GM/c^2$. Probably the time distortions are the hardest to grasp.

Exercise 12   In Unit 12, time measurements made by three different observers are compared. What are the positions and motions of these observers in relation to the black hole, and how are their time measures related? Why is Eddington–Finkelstein time introduced as a further time measure?

The event horizon of a Schwarzschild black hole is surrounded by another significant surface, the photon-sphere with radius $1.5k$. Light can orbit the black hole endlessly (though unstably) at this radius, and any light passing through the photon-sphere from outside is captured by the black hole.

Exercise 13   (a) Why would it be possible for a star such as a neutron star to possess a photon-sphere?

* (b) Could such a star be seen by its own light or by reflected light? How would it appear? What would it be like if you were on the surface of a star of radius exactly $1.5k$, assuming you could somehow survive?

A further property of black holes or near-black holes is the extremely strong tidal force that they can generate. Remember that the tidal force is the unmistakable signature of the curvature of space — it cannot be generated in flat spacetime even by an accelerating reference frame.

Exercise 14   Light of wavelength 600×10^{-9} m is emitted from a distant star radially towards a static black hole, and is received by an observer who maintains a fixed position at the radius of the photon-sphere of the black hole. What wavelength does this observer measure? If the Schwarzschild radial coordinate of the observer is 10^8 m, what is the mass of the black hole?

Exercise 15   How does the tidal force at the event horizon of a black hole depend on the mass of the black hole? What are the consequences?

4 Comprehension exercises for Block 4

1 The spectra of stars contain lines at well-defined frequencies. Name three sources of DOPPLER BROADENING of these lines.

2 Spectra of stars and galaxies are often shifted with respect to similar spectra observed on Earth. Name three possible mechanisms for such shifts, stating whether they would produce REDSHIFT or blueshift or both.

3 How is the LUMINOSITY of a star or galaxy related to its FLUX DENSITY and distance?

4 Define REDSHIFT in terms of observed and emitted wavelengths.

5 State HUBBLE'S LAW, and comment on its implications for our position in the Universe, and for the past state of the Universe.

6 What are the characteristics of THERMAL RADIATION, i.e. electromagnetic radiation inside a cavity with uniform-temperature walls?

7 Has any essential difference been found between the COSMIC MICROWAVE BACKGROUND RADIATION (corrected for local motions) and THERMAL RADIATION of the appropriate temperature?

8 The observed microwave background radiation is not in equilibrium with most visible matter in the Universe. Explain what this implies about the early Universe, with reference to DECOUPLING.

9 Why is NUCLEAR SYNTHESIS confined to a relatively narrow range of temperatures?

10 What are the principal differences between the conditions of NUCLEAR SYNTHESIS following the big bang and those currently occurring in stars?

11 State the COSMOLOGICAL PRINCIPLE. What experimental evidence supports it most strongly?

12 How does the cosmological principle allow us to infer a UNIVERSAL TIME?

13 Explain the term COMOVING COORDINATES.

14 What quantity in the ROBERTSON–WALKER METRIC describes the evolution of the Universe in time?

15 What quantity in the ROBERTSON–WALKER METRIC allows the spatial geometry of the Universe to be non-Euclidean?

16 Express the HUBBLE PARAMETER in terms of quantities occurring in the Robertson–Walker metric.

17 The simplifying assumptions of cosmology reduce the *source* terms in the FRIEDMANN EQUATIONS to a dependence on two quantities only. What are these quantities?

18 How does the energy density vary with scale factor in a MATTER-DOMINATED Universe?

19 How does the energy density vary with scale factor in a RADIATION-DOMINATED Universe?

20 How does the energy density vary with scale factor in a DARK-ENERGY-DOMINATED Universe?

21 What do the Friedmann equations predict for the sign of the DECELERATION PARAMETER q_0?

22 What are the special features of a CRITICAL UNIVERSE?

23 What is the astronomical evidence for the existence of DARK ENERGY?

24 What is the astronomical evidence for the existence of DARK MATTER?

25 What is the value of the DENSITY PARAMETER for matter, and what kinds of matter share this value?

26 What cosmological problems are resolved by the theory of INFLATION?

27 What source of energy is postulated to drive INFLATION?

28 Why do photons with wavelengths less than the PLANCK LENGTH presumably not exist?

29 What does SPONTANEOUS SYMMETRY-BREAKING have to do with inflation and the creation of matter particles?

5 Consolidation exercises for Block 4

Our explorations of space and time have brought us to realize that only general relativity can give us a detailed explanation of astronomical observations. However, leaving aside black holes, the distinctive effects of general relativity are quite slight on the scales of planets, stars, galaxies, and even clusters of galaxies. Newtonian gravitation serves very well; indeed it is still indispensable for much astronomical work.

Only on the scale of the distribution of clusters of galaxies does general relativity become essential. But to apply it we need to use a model in which the sources of curvature have high symmetry — Einstein's field equations being virtually insoluble for an arbitrary source distribution. In Block 3 we looked at the Schwarzschild metric, the solution outside a spherically symmetric mass distribution. In the final block, we examine the field equations and their solutions for a source distribution that is uniform throughout space. The assumption of homogeneous and isotropic space is known as the cosmological principle, and leads to a metric of spacetime on the cosmological scale named after Robertson and Walker. Insertion of the Robertson–Walker metric into Einstein's field equations leads to the Friedmann equations. Modern cosmology is based on the application of the Friedmann equations to the Universe.

5.1 Unit 13

The detailed arguments for the cosmological principle are postponed to Unit 14, but clearly we need to measure distances to remote objects and so Unit 13 begins by explaining how the astronomical distance scale has been established. A spatially uniform source distribution has a very distinctive feature — it is intrinsically dynamic. So cosmologists can use the Friedmann equations to look both backwards and forwards in time, provided we know the dynamics of the Universe now. Unit 13 also describes the extent to which astronomy can provide this information. We find that the Universe is expanding. Extrapolating backwards, it must have been denser in the past. We also find evidence of extremely high densities (and temperatures) in the early Universe from the microwave background radiation, and from the relative abundances of certain light nuclei.

These three areas of experimental evidence independently suggest that the Universe was once very much more dense and hot than we see it today. Taken together, they tell us that it underwent an explosive expansion at a very early stage. We call this expansion (and the original singularity from which the expansion grew) the big bang.

The first piece of evidence is Hubble's law relating the recessional speed of galaxies to their distance.

Exercise 16 State three methods of determining distances to galaxies. Briefly describe any restrictions on their applicability.

Hubble himself did not measure redshifts of more than $z = 4 \times 10^{-3}$ which corresponds to a distance of 11 Mpc or about 35×10^6 light-years. Now we are measuring redshifts of over $z = 6$.

Exercise 17   Taking the present value of the Hubble parameter to be $2.3 \times 10^{-18}\,\mathrm{s}^{-1}$, what cosmological redshift would you expect from a galaxy 5×10^8 light-years distant? If we measure a particular spectral line from the galaxy to have a wavelength of 504 nm, what would laboratory measurements, made in the galaxy, give for the wavelength of this line?

Exercise 18   Can redshifts as large as $z \geqslant 6$ be understood in the context of Equation 4? What does this imply for Equation 5? Revise Equation 4 to be in a form suitable for large redshifts, and calculate the recessional speed of a galaxy with a redshift of 2.

The existence and properties of the cosmic microwave background radiation provide the next pieces of evidence for the big bang. We cannot account for either its temperature or its isotropy assuming it to be in equilibrium with the matter that we know about in the present Universe. But if the early Universe was much denser, the background radiation would have been correspondingly hotter, and before getting 'decoupled', would have been in thermal equilibrium with hot ionized gas.

Exercise 19   Taking the typical present frequency of the cosmic microwave background radiation as $2 \times 10^{11}\,\mathrm{Hz}$, estimate the typical wavelength of the radiation at the time of decoupling. If there had been creatures then with eyes like ours, would they have been able to see it?

[Take the temperature at decoupling to be 3000 K.]

The term 'background' is peculiarly apt. Not only does this radiation reach us uniformly from all directions, but in a sense it provides a backcloth in front of which parade all visible galaxies and stars formed more recently, since decoupling. Looking to great distances is the same as looking back in time so we cannot see back to an earlier era than that of decoupling, since prior to that the Universe was opaque to electromagnetic waves. So, instead of seeing nothing beyond the most ancient galaxies, we find ourselves surrounded in all directions by this early, dense Universe! This is a modern version of the resolution of Olber's paradox, whereby the night sky in an infinite universe should be as bright as the Sun everywhere since a line drawn out in any direction would eventually meet the surface of a star. In fact the sky is extremely bright in all directions, but only to those having microwave sensitive eyes.

The evidence examined in Section 7 of Unit 13 refers to an earlier era still, when, with temperatures around $10^9\,\mathrm{K}$, light nuclei such as helium and deuterium were synthesized. The resulting proportions of these very light nuclei provide crucial support for our understanding of this early era of the Universe and imply limits on the present-day density of baryonic matter. The argument is based on modelling which relies on experimental data from both laboratory-based and astrophysical measurements.

Exercise 20   Section 7 refers to or implies experimental data of various kinds. List these, stating in each case whether the data are obtained by laboratory-based or astrophysical observations.

5.2 Unit 14

This Unit explores a model metric for the Universe, the Robertson–Walker metric. Einstein's field equations are satisfied for this metric, as long as the source terms (energy density and pressure) are isotropic and homogeneous. The Unit begins by reviewing the arguments for believing that the Universe is indeed isotropic and homogeneous, on a large enough scale.

Exercise 21 Is there *direct* evidence that the Universe is isotropic and homogeneous *now*? Explain with reference to different astronomical scales.

Despite their limitations, the isotropy data are quite convincing. The direct evidence for homogeneity of matter is not as strong as we would like, however (largely because of our ignorance of galactic evolution). The most overwhelming evidence for the cosmological principle comes from the cosmic microwave background radiation. This is fortunate, since the simplicity of the Robertson–Walker metric depends on it.

* **Exercise 22** Assuming that the Universe is spatially flat but expanding, explain why the gradient of the graph in Figure 6 indicates (for the sources of greater received power) a fall-off faster than $S^{-3/2}$.

The Robertson–Walker metric (Equation 2) has two coordinates with remarkable features. The radial coordinate σ is comoving, so that any observer, located in any galactic cluster, will never pass any other cluster. Also the time t is universal so that the clocks of all such observers, once synchronized, will stay synchronized despite the spacetime curvature. However, an observer who moves between clusters will experience a different time.

In Section 5 of Unit 14, Hubble's law is interpreted in the framework of the Robertson–Walker metric. Redshift is seen as being caused by space expanding and stretching the wavelength of light during the period of its travel from one cluster to another, rather than by a Doppler shift due to the recessional velocity of the source galaxy at the time of emission. The effect is characterized by the Hubble parameter H which is no longer regarded as constant in time, and by a deceleration parameter q.

Exercise 23 SAQ 10(a) indicates that if the Hubble parameter is in fact constant, then $q = -1$ and this implies accelerating expansion. Interpret this by considering the motion of a cluster. Also show that if $R = At^n$ where A and n are constant, $n \neq 0$, then $q = 1/n - 1$.

Of course, special relativity holds true in local inertial frames, and in such frames light has speed c. However, when light propagates over cosmic distances, in an expanding universe, speed of light c in one local inertial frame translates into a different speed in another frame; indeed, one has to be careful how one defines speed and in the following example it is done in terms of proper distance.

Recall that subscript 0 always
indicates the present era. See
the marginal note on p. 8 of
Unit 13.

Exercise 24   Write down the Robertson–Walker metric for a critical
universe. Substitute the form of $R(t)$ appropriate if the universe is
matter-dominated. Now consider events $\mathscr{E}_1$ and $\mathscr{E}_0$ at which a light pulse is
respectively emitted and absorbed. Use the metric to show that if by the
speed of light v in this universe we mean (the proper distance between the
events) divided by (the universal time interval $t_0 - t_1$), then
$v/c = 3(t_0 - t_0^{2/3}t_1^{1/3})/(t_0 - t_1)$ when $\mathscr{E}_1$ occurs at the origin and
$v/c = 3(t_0^{1/3}t_1^{2/3} - t_1)/(t_0 - t_1)$ when $\mathscr{E}_0$ occurs at the origin. Evaluate this
ratio for an observer receiving light emitted when the universe was half its
present age, and for an observer who sent light out when the universe was
half its present age, this light only now reaching a distant galaxy.

5.3   Unit 15

The Robertson–Walker metric on its own gives no information on the form
of $R(t)$ or the values of k or q. These spacetime parameters ultimately
depend on whatever conditions at the big bang determined the density of
the Universe. Einstein's field equations (of which there are ten in general,
corresponding to the ten components of Ricci curvature) lead to just two
equations when the source distribution is both isotropic and homogeneous.
These are the Friedmann Equations 1 and 2. They tell us the behaviour of
the curvature parameters provided we know what the source terms are.

Exercise 25   For a spatially flat, radiation-dominated universe
(a) Show that $R = At^{1/2}$ satisfies Equation 5.
(b) From Equation 1, show that the pressure p at time t is given by

$$p = \frac{c^2}{32\pi G}\frac{1}{t^2}.$$

(c) What will the deceleration parameter q be?
(d) Find an expression for the age t of such a radiation-dominated universe
in terms of $H(t)$, the Hubble parameter for the universe.

Recent observational evidence regarding our Universe, particularly Hubble
plots based on Type Ia supernovae and theoretical 'fits' to microwave
anisotropy data (WMAP etc.), indicate dark energy is currently dominant.

The relevant proportions are indicated by the density parameters which,
judging from various sources, seem to have the following approximate
values at present

$\Omega_{\Lambda,0} = 0.73$, $\Omega_{\text{matter},0} = 0.27$, $\Omega_{\text{radiation},0} = 0.00005$.

Baryonic matter accounts for about one-sixth of the matter, so $\Omega_{\text{b}} = 0.044$,
in keeping with our expectations based on big bang nuclear synthesis.

The total energy density parameter is very close to 1. If $\Omega_0 = 1$ exactly,
then the Universe has a flat spatial geometry and the total energy density
will have the critical value:

$$\rho^c = \frac{3H^2c^2}{8\pi G}.$$

Given that current estimates of the Hubble parameter indicate
$H_0 = H(t_0) = 2.3 \times 10^{-18}$ s^{-1} (about 71 km s^{-1} Mpc^{-1}), it follows that the
total mass density of the Universe is now $\rho_0^c = 1 \times 10^{-26}$ kg m^{-3}. This is
consistent with a Universe that has a current age of 13.7×10^9 years and

which became transparent when decoupling occurred about 380 000 years after the big bang.

What we have so far loosely referred to as the 'big bang' was the period of nuclear synthesis, when the Universe was a few seconds old or about 10^{-17} of its present age. In Section 6, we try to look back to an era occurring before 10^{-30} of its age *then*. The theory of inflation may seem audacious, but what else could be expected when we are dealing with energies well beyond what we could ever achieve in accelerators? We do not yet have a satisfactory theory of quantum gravity, but the basic energy driving the inflation, the energy of quantum fluctuations in a vacuum, is well established theoretically and, moreover, has experimental consequences that have been verified to extraordinary accuracy.

Exercise 26   Assuming that the inflationary period lasted 10^{-32} s and that the energy density then consisted only of the density of vacuum fluctuations of $10^{94}\,\mathrm{J\,m^{-3}}$, calculate, using Equation 13, the factor by which the Universe expanded during inflation. [Hint: Use the result of SAQ 25.]

6   Review and synthesis of the whole course

Abbreviations used in this
review: NM, SR, GR, CS.

We now look back over all four Blocks and trace some of the key ideas of
the course as they develop. Some notions have been abandoned along the
way, others modified or generalized. Blocks 1–3 represent three quite
separate 'world-views' of the Universe in terms of events in space and time.
Block 4 retains the world-view of the general relativity of Block 3, yet
because it deals with a unique system — the Universe itself — has
sufficient special features to deserve equal representation with the other
three. For brevity we write NM (Newtonian mechanics), SR (special
relativity), GR (general relativity) and CS (cosmological scale). 'Scale'
serves as a reminder that in CS we are not concerned with the detailed
structure of galaxies, clusters, etc. but with a broad overall description.

The assumptions A1 to A12 of Unit 1

All 12 assumptions are implicit in NM. How many have survived? We
discuss them here in the order in which they appear in Unit 1.

A1   The passive nature of space and time

***The properties of space and time are not affected by external
influences, such as the distribution of matter or energy.***

A1 is implicit in SR as well as NM. The properties of space and time are
revealed through the study of the motion of particles and light rays. In
NM and SR, future motion depends on two things only — present motion
and the effect of forces. The time or space in which the particle moves
plays no part except to allow the motion to be measured. However, in GR
and CS, that motion which is due to gravity alone is determined by
present motion and the geometry of the spacetime in the neighbourhood of
the particle (or photon). This geometry is altered by 'sources', i.e.
distributions of mass, energy, momentum and momentum flux. Within the
sources, the so-called 'Ricci curvature' of the spacetime is non-zero,
according to Einstein's field equations. Outside the sources, where the
Ricci curvature is zero, the continuous change in the Riemann curvature,
governed by Einstein's field equations, ensures that the 'sources' modify
the geometry everywhere. The field equations ensure that the effect falls
off with distance from the sources, exactly as we expect for a gravitational
effect. In CS, viewing things on a scale where the cosmological principle
holds, all regions contain an identical source density at a given time, so one
can never be remote from sources — there are no 'external influences'.
However, the source density changes with time as the Universe expands,
and the spacetime geometry and consequent motion of light and particles
(clusters of galaxies) change accordingly.

A2   The homogeneity of space

The properties of space are the same in all parts of the Universe.

Again this is implicit in NM and SR. However, in GR, the properties of space are influenced by source distributions near the region of space in question. In particular, GR tells us that spatial geometry is non-Euclidean. Since matter, energy and momentum are not uniformly distributed in the Universe, it follows that space itself is inhomogeneous. We only expect dramatic departures from Euclidean geometry close to black holes or near black holes, but all effects ascribed to gravity in NM are associated with curved space (and, more particularly, curved spacetime) in GR. In CS, the source distribution is uniform so space is homogeneous. This does not necessarily mean that it is flat; that would only be the case if the spatial curvature parameter k has the value zero. Any curvature that the space of the Universe has would change over time as the source density alters with expansion because the spatial curvature of the Universe is k/R^2.

A3   The continuity of space

Between any two points in space, we can find another point.

This assumption is implicit in all four Blocks although in Block 3 (Unit 12) we met the Planck length, which places a lower limit on the possible wavelength of light. Since we have no theory of quantum gravity as yet, the full implications of this are not really known.

A4   The isotropy of space

The properties of space are the same in all directions.

The situation is much the same as for A2. Isotropy of space is implicit in NM and SR, but in GR space is not isotropic because the distribution of sources is manifestly anisotropic. However, for CS the source distribution is isotropic, thus space is also. This only applies, though, to observations made on the scale of clusters of galaxies, by an observer riding a cluster. For an observer moving between clusters, the mean distance between clusters will be reduced in the direction of motion, but unaltered in the perpendicular directions as described by the Lorentz contraction of SR. Hence for such an observer the source distribution is anisotropic and so the space will be anisotropic too.

A5   The completeness of space

Each ordered set of three real numbers (x_P^1, x_P^2, x_P^3) is associated with a unique point P in space.

We have met nothing to discredit this assumption. However, some physicists take seriously suggestions that there may be 'wormholes' in spacetime giving 'short-cuts' to other regions and eras of the Universe. On a different scale, there have for years been theories that spacetime has many more than four dimensions, with the 'spare' dimensions tightly rolled up.

A6   The homogeneity of time

All instants in time are equivalent.

This means that identically set up instances of an isolated experiment give the same result every time it is repeated. In NM and SR we assume that A6 always holds in *inertial reference frames* — we would scarcely expect experiments performed in the back of a van being driven along a winding road to yield repeatable results.

In GR we can notionally set up experiments far from curvature sources, (or include a source such as a planet within the experiment) so A6 still holds. However, on CS we can never escape the pervasive influence of the expanding Universe and A6 fails.

Time on CS is characterized by a unique event, the big bang. Cosmologists commonly regard time as well as space to have originated with the big bang, so whether or not time will continue infinitely into the future, its past is finite, at least for this Universe.

A7   The continuity of time

Between any two instants in time, we can find another instant.

This is implicit throughout the course. As with length, there is a proviso connected with quantum gravity (Planck time).

A8   The independence of space and time

Space and time are quite independent concepts.

In Unit 1, 'independent concepts' in this context was taken to mean that spatial properties do not change with time, and that time-scales are universal. In other words, a ruler has the same length at all times, and a clock ticks at the same rate everywhere. In SR we met time dilation and length contraction which affects rulers and clocks at rest in one inertial frame when observed from another in relative motion. But A8 still holds *within any one inertial frame in which the rulers and clocks are fixed.* Only when transforming between inertial frames in relative motion do space and time 'mingle' in the Lorentz transformations, but then they also mingle in the Galilean transformation $x' = x - vt$.

However, there is an important sense in which space and time lose *some* of their independence in SR; this could not have been anticipated in Unit 1. Consider first a particle moving parallel to the x^1-axis; its x^2-coordinate remains constant whatever the value of x^1. We say that x^2 is independent of x^1 and of course the converse is also true. But if we transform to another frame by static rotation of say 30° about the x^3-axis, then the same motion will appear as changes in *both* x'^1 and x'^2. The independence of the orthogonal coordinates is apparently lost. Clearly this is because (x^1, x^2) denotes a two-dimensional space in which any pair of orthogonal axes may be drawn. The Lorentz transformations involve a mixing of space and time for a similar reason: they represent a sort of rotation in the four-dimensional Minkowski spacetime (x^0, x^1, x^2, x^3). Space and time become welded together.

But spacetime is not isotropic; we cannot draw orthogonal (ct', x'^1) axes in any direction because the relative velocity of two inertial frames must be less than the speed of light. The time axis is a special direction. Time proceeds only forwards and $(\Delta t)^2$ enters into the expression for the Lorentz invariant interval as a positive quantity, while the spatial terms are negative. We use different instruments to measure space and time. So, despite the usefulness of Minkowski spacetime as an abstraction, we have to recognize that ct and x are different in a way that x^1 and x^2 are not. Space and time remain separate concepts.

In the curved spacetime of GR, space and time continue to be distinct in this sense. But the Unit 1 meaning of A8 is no longer tenable.

A clock at the top of a tower ticks faster than a clock on the ground, even though both are at rest relative to the planet they are on. However, neither of these clocks is at rest in an inertial frame since they are not in free fall. Even if two clocks are in free fall, and are sufficiently separated for the difference between their rates to be measurable, they could not both remain at rest in the same inertial frame. Time is dependent on space because inertial frames are only local.

Equally, spatial properties are in general time-dependent. For example, a triangle in a frame projected outwards from a dense star will have angles whose sum will progressively approach 180° as it enters regions of progressively smaller spatial curvature. Abandonment of global inertial frames has caused space to alter time and vice versa, if we insist on using a reference frame that is too big to be inertial.

The cosmological redshift provides a further twist to this story. If treated by NM or SR, it is accounted for by the classical or relativistic Doppler effect. Neither challenges the independence of time from space; according to SR, the clock on a distant cluster is moving so it ticks more slowly than ours. The effect is reciprocal in that a distant observer would say our clock is slow. But GR tells us that cosmological spacetime is curved, so the inertial frame for one cluster is only local and cannot contain a distant cluster. Moreover, according to the Robertson–Walker metric, with its comoving coordinates, we regard *both* clocks as being at rest, even though the proper distance between them is changing! Like all clocks at fixed Robertson–Walker coordinates, they keep the same universal time.

For clocks recording universal time to stay synchronized, they have to have constant Robertson–Walker coordinates. Clocks separated by constant proper distance cannot remain synchronized. So in saying universal time is independent of space, we specify Robertson–Walker space not 'ruler' space.

So is space independent of time? Clearly it is, if we use the Robertson–Walker coordinates (σ, θ, ϕ) as a measure of space, because clusters retain their coordinates indefinitely as time progresses. But the proper distance between clusters, which is the distance that would be measured by rulers placed end to end, increases because of the time-dependence of the scale factor $R(t)$.

CS also allows space to have a curved geometry. The curvature is $k/(R(t))^2$, a function of time unless $k = 0$. The analogy of the inflating balloon shows that, for example, the ratio of circumference to diameter of a circle increases with time towards π. In this sense too, CS space does depend on time.

A9   The universality of time

All observers should agree on the time interval between two given events. In particular, they can agree whether or not two events are simultaneous.

A10   The universality of distance

All observers should agree on the distance between two simultaneous events.

Both of these basic assumptions of NM had to be abandoned when we moved to SR. If two inertial observers who use identically calibrated instruments are in relative motion, they nevertheless agree on the value of the speed of light. A9 and A10 then become impossible to sustain. Two events $\mathcal{E}_1$ and $\mathcal{E}_2$ that are simultaneous for one observer (because, for example, they are both triggered by light from an event $\mathcal{E}_0$ which is equidistant from the sites of both $\mathcal{E}_1$ and $\mathcal{E}_2$) cannot be simultaneous for the other observer.

In GR, the failure of A9 and A10 was extended to two separated observers who are not in relative motion, although it became impossible to locate both of them in the same inertial frame.

In CS, we turned full circle and recovered a 'universal' time-scale, applicable to all observers at rest relative to nearby clusters (and not significantly perturbed by nearby massive objects). Although a spaceship travelling between clusters would have a different proper time, the two time-scales are not equivalent and the universal scale is the more appropriate for describing CS events. To the extent that different observers can agree that two widely separated events are simultaneous, they can also agree on the distance between them.

A11   The invariance of the distance function

If two simultaneous events occur at points P *and* Q, *the value of the expression*

$$|PQ| \equiv \sqrt{(x_Q^1 - x_P^1)^2 + (x_Q^2 - x_P^2)^2 + (x_Q^3 - x_P^3)^2}$$

is unaffected by translations, rotations, reflections or boosts.

This is an assumption about space as described in a particular coordinate system — the Cartesian system. It is true only when space is flat, as it is in NM. As we have progressed through the course, it has become transmuted into a concept of central importance: the metric of spacetime. Let us review the steps in this process.

1 First we note that while A11 characterizes the spacetime of NM, it must be abandoned in SR for linear boosts, because as noted under A9, simultaneity of separated events is lost under this transformation.

2 Next we found that in the spacetime of SR, there is a function of the coordinates of two *events* P and Q which *is* invariant under all transformations between equally calibrated inertial frames; it is the Lorentz invariant interval S_{PQ} and is a sort of four-dimensional distance function although it is only real (i.e. $|PQ|^2$ is positive) for time-like rather than space-like intervals.

3 In Blocks 3 and 4, we explored other coordinate systems, so we need to re-express A11 in generalized spatial coordinates. Immediately, a problem arises. We can write an expression for a small distance Δl in terms of small changes in generalized coordinates Δq^1, Δq^2, Δq^3. This expression is called a metric and has metric coefficients which are functions of the coordinates. To use the metric to obtain a distance function between general points, we need to know two things: how the metric coefficients depend on the coordinates, and the path along which to integrate. In the

Cartesian system, the metric coefficients are all equal to one when the space is flat, and the function quoted in A11 automatically gives the shortest distance. In other systems, metric coefficients may vary and the path must be specified, e.g. by showing how the coordinates vary with a parameter appropriate to the path. There may be no simple formula for the length of the shortest path.

4 The metric formulation gave us the opportunity to define a curved space, i.e. a space with geometric properties different from those of Euclidean space. We were able to define the curvature of a two-dimensional space in terms of its metric coefficients.

5 The metric of spacetime is an expression for $\Delta S/c = \Delta\tau$ in terms of generalized coordinates; its most general form includes ten metric coefficients.

6 GR asserts that space and spacetime really are curved. The metric of spacetime reduces to the SR expression for $\Delta S/c$ only in a freely falling local spacetime region. The variation of the metric coefficients which produces Riemann curvature produces measurable consequences on a more extended scale. In its original form, A11 provides the answer to the problem of finding the shortest route between two points — it is a straight line. In GR, the equivalent problem is finding the world-line of maximum proper time between events. Such a world-line solves the geodesic equations for the metric, and it is a central postulate of GR that it represents a possible world-line of a particle or photon in free fall. Remember that photons *always* travel on null geodesics, i.e. geodesics for which the interval of proper time is zero.

A12

There exists a triangle in which the sum of the angles is 180°.

A12'

There exist two triangles with equal angles but unequal sides.

These postulates were self-evident to Newton and the Euclidean geometry characterized by them is the spatial arena for NM. According to SR, the linear boost transformation does strange things to triangles. If the boost is in a direction lying in the plane of the triangle, lengths with a component in that direction will be altered, and so in general will angles. However, A12 and A12' remain true in each inertial frame.

According to GR, spatial geometry is non-Euclidean and no triangles exist whose angles sum to exactly 180°. Of course, except near black holes the departures from the Euclidean result are extremely small. However, strict Euclidean geometry in GR only applies in the limit of infinitesimally small figures in free fall. The departure from the sum of 180° increases with the area of the triangle, so A12' is just as invalid as A12.

In CS, spatial geometry is flat if the spatial curvature parameter $k = 0$ as is widely believed. A12 and A12' would then be true on the largest scales, but CS is inapplicable on scales of only a few clusters. The situation is almost the reverse of that in any other GR application!

The principle of relativity and inertial frames

A13 was a preliminary statement of A14.

We start with the comprehensive statement of Galilean relativity, A14 of Unit 3. This refers to two inertial observers, i.e. observers using inertial frames of reference. The criteria for checking that a particular frame is inertial are set out in Section 5 of Unit 2. They make use of four 'free' particles: one to mark the origin, the others to define mutually perpendicular, non-rotating axes. The problem, of course, is ensuring that the particles are free from any net force. But our intuition suggests that forces between objects decrease as their separation increases. Thus if we take our particles 'very far' out into space, the gravitational influence of the Earth, Sun, etc. might be negligible. This works for NM and SR, but is only accurate, we now realize, for 'empty' spacetime (and NM only for low velocities). Our spacetime, according to the cosmological principle, is by no means empty. Even if we escape the gravity of a cluster of galaxies, we will eventually be affected by other clusters. So, in terms of absolutely precise measurements, we conclude that there is nowhere to go to test whether our frame is inertial or not.

Inertial frames involve another assumption: once the position of the origin and the axis directions are established, the axes can be extended indefinitely. But sooner or later a planet, star, or other source of gravitational influence would be encountered. Test particles projected into their neighbourhood would not travel with constant velocity because, according to NM and SR, they are no longer free. But GR asserts that they are free — in free fall — until acted on by non-gravitational forces. Thus the failure of N1 is due to the fact that an inertial frame can *not* be indefinitely extended, as assumed in SR. A particle in free fall, according to GR, passes through a succession of inertial frames, each of them applicable only within an infinitesimal spacetime volume.

The principle of relativity (A14) is a statement about experiments conducted by inertial observers who each have an isolated system of particles. How do we ensure a system is isolated? The problem of isolation that we had when we tried to define an inertial frame precisely, has recurred. We could sidestep it as before, by taking the experiments deep into space and neglecting residual gravity there. Or, since the particles *within* the system may interact with one another, we could include the Earth as one of the particles, and remain on its surface. (Of course to do this we would need to neglect the Earth's rotation, orbital motion, and the effects of other heavenly bodies.) A14 serves well in NM and SR whether the interactions between particles are gravitational or otherwise.

Within GR, we have a problem because, as we saw in connection with A2, A4 and A6, space and time are no longer homogeneous, nor space isotropic once gravity is seen as a geometric effect. So the results of an 'isolated' dropping experiment would depend very much on the position and orientation of the axes used. In a boosted frame of reference, the results would depend on the boost speed and on the time that the experiment was started. Of course all such frames would fail the test for being inertial; it is not that A14 fails ... it doesn't apply.

So what do we do with A14? We can restate it with minor but crucial changes. We confine both observers, and the experiments they observe, to *local* inertial frames. (The two frames in the statement of A14, however, may be widely separated in space and/or time.) The test particles may be near planets or stars and will be in free fall. Then, in the limit that their frames have 'small enough' spacetime volumes, the differences will be too small to be detectable by the measuring equipment of the observers. The expression 'freely falling frames are locally inertial frames' implies that such frames exist in every spacetime region and that A14 applies, with appropriate accuracy, within each of them. This is the key to generalizing special relativity to produce a theory of gravitation.

The new, local, form of A14 applies quite generally but the size (four-dimensional volume) of a local frame depends on context. The spacetime volume of a free-falling inertial frame close to, say, a neutron star, would have to be very much smaller than that of a frame in intergalactic spacetime.

In principle, one could envisage the real universe as composed of a mosaic of such frames of appropriate sizes, within each of which the modified A14 applies. But this would be mathematically intractable and could not be used for cosmological modelling. Instead, we assert that, on large enough scales (denoted by CS), the Universe is homogeneous. We have a mathematical model which embodies this, and we can try to test A14 in this model, thus leaving aside any problems of local 'lumpiness'. It turns out that as far as most transformations are concerned, the principle of relativity applies without restriction to local frames. Because space in CS is homogeneous, the frames can be of any spatial extent, and A14 applies in respect of all the static spatial transformations. We can choose any cluster of galaxies as an origin for Robertson–Walker coordinates, and also orient and reflect axes as we wish. Note, however, that CS time is inhomogeneous because of the acceleration or deceleration of the Universe.

In our critical examination of how A14 survives the world-view of GR, we have overlooked its role in the intermediate case of SR. Unit 5 Section 3.3 contains the two postulates of Einstein's 1905 paper, SR1 and SR2. SR1 is a concise restatement of A14. SR2 asserts the invariance of the speed of light in free space, which is equivalent to stating that SR1 applies to Maxwell's equations. SR1 is a sort of 'law about laws' without specific content, and the fact that SR1 applies to Maxwell's equations is in conflict with Newtonian physics and thus far from obvious. So, SR2 is of central importance in replacing the world-view of NM by that of SR and Einstein justifiably gave it equal prominence with SR1. When Maxwell predicted that electromagnetic waves would travel at about $3 \times 10^8 \, \mathrm{m \, s^{-1}}$, he and his immediate successors believed that a pervasive 'ether' was the medium in which the waves travel. Michelson later showed that the measured speed of light was independent of the motion of the laboratory through this supposed medium. There was now no way that SR1 and SR2 could be reconciled with the spacetime of NM and many physicists were prepared to abandon the relativity principle. Einstein reaffirmed the principle but in doing so overturned many of the pillars of NM, notably the universal time-scale.

Einstein's assumptions SR1 and SR2, as quoted in Unit 5, make no statement about the size of inertial frames or systems. It must be emphasized that they only survive the extension to GR when that restriction to local inertial frames is introduced. If we know the metric of curved spacetime, we can 'determine the speed of light' over a large region by calculating the proper length it travels in a given coordinate time interval, using the fact that the proper time interval between any pair of events linked by a flash of light is zero. This was done for example in Section 2.3 of Unit 12; we saw how light emitted from objects approaching the event horizon of a black hole is characterized as 'slowed down', but in doing so it has traversed innumerable different inertial frames before being detected by a distant observer. In the case of CS, the expansion of space allows light to travel either *faster* or *slower* than c, depending on whether the observer is emitting or receiving the light. This effect, which is appreciable over cosmological distances, was shown in Exercise 24 of Section 5.2. But within a *local* freely falling inertial frame, the speed of light remains c.

7 Essay exercises for Blocks 1–4

1 Write an essay on the developments of the ideas of dynamics since the time of Newton. It should include discussion of the following:

Simultaneity and the relationship between space and time;

The concepts of mass and energy and their interrelation;

The effect of matter on the geometry of space and time;

The space and time of cosmology.

2 Three very different physical theories are commonly in use today: Newtonian mechanics, special relativity, and general relativity. Write an essay to explain how this comes to be the case. In particular, address the following points:

The hierarchical relationship between the theories, and the range of conditions over which it is appropriate to use each of them;

Two observational tests that refute Newtonian mechanics in favour of special relativity;

Two observational tests that refute Newtonian mechanics in favour of general relativity.

8 A tutor's perspective — a personal overview

Opinions expressed in Section 8 are not necessarily those of the S357 Course Team.

This final section is one of personal reactions: yours as a current student and mine as a tutor whose outlook has been enriched by reactions of many previous students. **None of the material in this section is assessable and if you want to read it you could easily wait till after the exam, when you can afford a more relaxed retrospective survey of the course.** So, for now, put aside any worries you may have concerning your own assessment, and be yourself an assessor of the course! By this we mean that you should perhaps try to evaluate the contribution the course has made to your own experience and the extent to which it has enhanced your overall study programme, possibly complementing or clarifying other courses you may have taken or are intending to take.

Students often find the course quite different to what they expected. A typical remark is that 'A new kind of thinking is needed'. Why should this be? Probably because, along with the anticipated mathematical and physical strands, there is a third, more philosophical, strand. Now historically, physics and philosophy were very closely allied; indeed older universities used to offer 'Natural Philosophy' rather than physics courses, until quite recently. But sadly, physics teaching in schools has become the presentation and assimilation of 'facts', and even in universities the need to produce graduates with sufficient current knowledge has eroded the teaching of the historical and philosophical roots. Many physics graduates are unaware of the existence of rival theories in prominent areas, and I would guess that many leave school thinking that experimental physics exists only to confirm theory. Now S357 makes us confront head-on the problems of the matching of 'reality' (as revealed by experiments) with a scientific description of it, and for many of us the necessarily philosophical thought processes involved are a shock, though they shouldn't be.

Perhaps an inevitable consequence of the teaching of scientific theories as 'facts' was a backlash from the sociologists. The claim that the scientific method gives us a unique handle on the 'truth' has prompted a counter-claim encapsulated in the phrase 'Science is a social construct'. The extreme adherents of this view regard Science as *nothing more than* a cultural phenomenon of Western society, in much the same way as they might regard fashion, law, morals, or music. Some have even used Einstein's relativity (taken as 'everything is relative') as a justification, pointing out how, for example, the view of which sexual practices are acceptable or unacceptable depends on the society one is considering.

There are two aspects of physical theories which S357 illustrates well. Both these aspects are difficult to explain to non-scientists. (Whether or not sociologists are scientists is a separate matter.) One aspect is the nature of 'scientific revolution'. As scientists, we use this phrase to signify a breakthrough or advance achieved by great scientific vision, requiring the abandonment of old thought moulds, and gaining general acceptance through improved or simplified explanatory and predictive power. However, to the non-scientist, it can appear as the replacement of one theory, previously thought to be 'true', by another. The old theory is 'discredited'. The advances from Newtonian spacetime to the flat spacetime of special relativity, and then to the curved spacetime of general relativity, provide excellent examples. The scientific perception is that in each case, the old theory survives as a special case of the new theory. It may be superseded but is not discredited. Sometimes an aspect of an old theory (simultaneity as absolute, gravity as a force) is denied in the new theory, but would still be used in appropriate circumstances. Each new theory owes a large debt to the old. Newton himself (and he was not particularly modest) admitted that, if he had seen further than other men, it was by standing on the shoulders of giants. This memorable metaphor portrays science both as a human endeavour and one that advances (not just changes) and truly explores the Universe.

The other aspect of physical theory illustrated by S357 is the internal structure of a theory as distinct from the nature of advances. This also throws light on the 'social construct' aspect of science. Science teaching has traditionally, and almost unavoidably, endowed every scientific theory with the same structure as classical mathematics or logic. Scientific 'laws', a relatively small number of very general statements, take the place of the axioms of mathematics. Then by mathematical and syllogistic progression, various other general principles can be 'proved'. Finally one can apply the fundamental laws or the secondary principles to the solution of particular problems.

But this pyramid structure is quite illusory. For instance, consider classical mechanics. It is true that the whole of this can be developed from Newton's laws of motion (and any required force laws); this approach is illustrated in Unit 2. However, Unit 3 develops classical mechanics from principles of symmetry and conservation laws. It is futile to argue over which is the 'correct' starting point. Newton's laws are perhaps more amenable to experiment and their assertion is at least justified on pedagogic grounds to counter the intuitive notion of motion, rather than acceleration, being caused by force. Energy is a more esoteric concept but one applicable outside mechanics. Moreover, these are only two of many starting points. In mathematical physics courses, classical mechanics is often taught starting from the Lagrange equations (which are written in 'generalized coordinates'). This approach is particularly useful in that compared with the Newtonian method, it leads much more easily into quantum mechanics. Yet another approach is by the 'least action' principle. Here a quantity is evaluated along various paths and the actual path is the one for which the quantity is minimized. You will see here a close analogy with the geodesic principle of general relativity; the same idea was used by Feynman to develop quantum electrodynamics.

The point we are making is that classical mechanics, and probably every other useful theory, does not have a pyramid or tree structure. Rather it resembles a string bag, rolled into a loose irregular ball. It is possible to start anywhere and visit the whole. The only essential item for the structure of any theory is that it must be *internally consistent*. (This means that if we start at one place and arrive at proposition P, we cannot also start from another place and arrive at a denial of P.) It is also important that all the concepts in a theory are *connected* by logical pathways. If we impose a simple structure on a theory, in particular so that we can learn it (you have to start *somewhere!*), then we can acknowledge that this is a social construct. But scientists would argue that the theory itself, as distinct from our representation of it, has a validity in relation to the real world, and that this is so regardless of whether or not the theory is the latest model. The word 'validity' rather than 'truth' has been used here since many physicists as well as sociologists would acknowledge that truth is a very slippery fish indeed and rarely either claimed or explicitly sought by theoreticians. However, the term 'validity' seems to cover the sorts of usefulness which scientists most value in practice (explanatory and predictive power, basis for technological design, fertility in respect of further theoretical development).

S357 is certainly a course about the various theories of space and time, and about the theories of mechanics and electromagnetism which are inextricably bound up with space and time. But it is also a course about *the relationships between theories*. (Many assignment or exam questions begin with phrases such as 'according to the Newtonian world-view ...' or 'write a reasoned comparison of the Newtonian and general relativistic descriptions of ...'. You will gather from the above discussion that answers along the lines of 'Einstein showed that Newton was wrong' will scarcely impress the examiners.) So it is appropriate to remind ourselves in more detail of the ways in which theories are judged.

The necessity of experimental testing is the feature which above all distinguishes science from other rational disciplines. A theory which is not susceptible to experimental tests is not, for most philosophers of science, a scientific theory. A theory that consistently fails tests is soon discarded. A theory that passes some tests but not others can be a valuable scientific tool provided its limits are recognized. A theory that passes all available tests has a good chance of being widely accepted in the scientific community. Of course no amount of experimental work can ever *prove* a theory, since there is always the possibility of alternative explanations. S357 provides many examples of crucial experimental tests.

However, science in general, and physics in particular, has never been satisfied with simply accumulating laws and formulae which seem to 'work'. There is a gut conviction that natural laws should be beautiful. This aesthetic requirement can be characterized in two ways: economy and simplicity. Historically, many theories have used *a priori* assumptions, often unrecognized as such like A1–A12 of Unit 1. We always prefer an *economical* theory, i.e. one using fewer assumptions, to one in which assumptions proliferate, given that both are experimentally sound. For example, the 'standard big bang' model assumes isotropy and homogeneity of the Universe, whereas 'inflationary' models, showing how this could arise naturally, are seen as advancing cosmological theory.

Ever since Newton united celestial and terrestrial mechanics, scientists have also sought to *simplify* their theoretical framework by finding underlying laws which apply in disparate realms. We have seen (Unit 4) how Maxwell united electric and magnetic field laws, and how this prompted Einstein to find a spacetime in which both electromagnetism and mechanics could operate. We also saw (Unit 9) how general relativity disposed of the distinction between inertial and gravitational mass. The vacuum energy fluctuations that drive inflation (Unit 15) are a consequence of quantum electrodynamics, the theory which unites electromagnetism and quantum mechanics. Finally we have pointed to the need, not yet fulfilled, to unite gravity with the other forces of nature.

Nearly all the theories and laws that we have examined are plainly testable experimentally. It is remarkable that we have been able to base our understanding of the early Universe, back to an age of perhaps 10^{-10} s, not simply on astronomical observations but on properties of matter and particles investigated in laboratories. However, Unit 15 inquires into times so much earlier that no manufactured accelerator could ever produce particles of the necessary energies. Rival theories (e.g. of spontaneous symmetry-breaking or incorporating quantum mechanics into general relativity) may never have observable consequences by which they could be tested. If we are denied experimental tests and can judge theories only by their aesthetic qualities, do we still have a science? Here is a frontier; whether we intend eventually to proceed or to turn back, it is a good place to stop.

9 Appendices 1–7

9.1 Appendix 1: Invariances (especially with respect to reflections and rotations)

In many cases, the correct answers can be found by either inspection or 'brute force'. For instance, the expression $|\mathbf{v}|^2/|\mathbf{a}|$ (where $\mathbf{v}$ and $\mathbf{a}$ are respectively velocity and acceleration) must be invariant with respect to origin shifts since inspection reveals only velocity and acceleration vectors, both of which are unaltered by adding an arbitrary constant vector to all position vectors. If all terms in the expression are unaffected, the expression itself is unaffected.

However, it does *not* follow that if all terms *are* affected, the expression is not invariant. To show this, we consider how the same expression transforms under a recalibration of time, using the 'brute force' method. In this method, we simply substitute transformed quantities into the expression and see what comes out. A recalibration of time is represented by $t' = \alpha t$. Since $\mathbf{v} = \mathrm{d}\mathbf{x}/\mathrm{d}t$ it follows that $\mathbf{v} = \alpha\,\mathrm{d}\mathbf{x}/\mathrm{d}t' = \alpha\,\mathrm{d}\mathbf{x}'/\mathrm{d}t'$ (since $\mathbf{x}' = \mathbf{x}$) so that $\mathbf{v} = \alpha\mathbf{v}'$. You can similarly show, from the definition $\mathbf{a} = \mathrm{d}^2\mathbf{x}/\mathrm{d}t^2$, that $\mathbf{a} = \alpha^2\mathbf{a}'$. So all quantities in the expression change under the transformation. But *the expression as a whole* remains invariant. Pursuing the 'brute force' method, you will find that $|\mathbf{v}'|^2/|\mathbf{a}'| = |\mathbf{v}|^2/|\mathbf{a}|$ because the velocity and acceleration occur in such a way that the factor α cancels.

If you are not confident with the inspection method, use the 'brute force' method. You could, for example, examine whether the same expression is invariant with respect to linear boosts or spatial recalibration.

If you are asked about reflections, they would be of one of the Cartesian axes. To use the 'brute force' method, you just reverse the appropriate components of vectors, and leave the remainder unchanged. For example, a reflection of the **2**-axis transforms the vector $\mathbf{p}$ by $p'^1 = p^1$, $p'^2 = -p^2$, $p'^3 = p^3$. Any vector perpendicular to the reflection axis is left unaltered.

Rotational transformations pose different problems. The 'brute force' method is applicable in principle, but for a rotation by an arbitrary angle about an arbitrary axis, it would mean substituting for each component of each vector in the expression a complicated linear combination of all three components in the new frame of reference. So forget it, and instead follow these simple rules:

1 All *scalar* expressions are invariant with respect to *all* rotations *provided they are independent of the coordinate system*.

Examples: the modulus of any vector, or any scalar product such as $\mathbf{v} \cdot \mathbf{a}$ or $p^1 x^1 + p^2 x^2 + p^3 x^3$ (even though this seems to refer to specific components). On the other hand an expression such as $p^2 x^1 + p^1 x^2 + p^3 x^3$ is not independent of the coordinate system.

2 All *vector* expressions are invariant with respect to rotations *about their own direction* but *not* about other directions. The same applies to any scalar expressions which could represent the component of a vector, for the direction of that component.

Examples: The vector $(0, 0, w)$ is invariant with respect to all rotations about the **3**-axis but not with respect to rotations about any other axis. The quantity $a^2 b^3 - a^3 b^2$ is the **1**-component of the vector $\mathbf{a} \times \mathbf{b}$ and so is invariant with respect to rotations about the **1**-axis. On the other hand, the scalar $a^1 b^3 + a^3 b^2$ is not invariant with respect to any rotations.

9.2 Appendix 2: Quantity invariance and form invariance

Invariance is a key word of the course, and refers to two separate concepts. You should always make clear in descriptive answers whether you are referring to form invariance or the invariance of the value of a quantity. The following notes are to remind you of the distinction.

In Appendix 1, we looked at various expressions to see whether they altered under particular transformations. If they did not, their numerical values in any specific case would not alter; the quantity they represent would be invariant under these transformations. But if they did, quantity invariance would be absent. Now these expressions might be for important dynamical quantities such as angular momentum or kinetic energy, or they could be fanciful constructs. It makes no difference which they are; each expression must be examined individually for each transformation. Kinetic energy, for instance, happens to be invariant with respect to all displacements, reflections and rotations but not with respect to linear boosts and recalibrations. It is possible to construct expressions with all manner of invariances.

Form invariance is a completely separate concept. Here we are concerned not with expressions but with *equations* which could represent *physical laws*. Whether or not the quantity represented by either side of an equation is unchanged in the transformed coordinates is immaterial; what we want to know is *whether the equation is unaltered* so that it can still be written (in the new reference frame) with the same arrangement of symbols. Consider, for example, the equation (for a system of two particles A and B) $m_A \mathbf{a}_A = 3(\mathbf{x}_A - \mathbf{x}_B)$. The quantity on each side of the equation is a vector and therefore not invariant with respect to general reflections or rotations. However, the *equation* still holds in any frame obtained from the first by reflection or rotation, that is $m_A' \mathbf{a}_A' = 3(\mathbf{x}_A' - \mathbf{x}_B')$, so the equation is *form invariant* under this transformation. (We assume that $m' = m$.) The vector quantity on each side of the equation would have different components in the new frame, compared with before the transformation, but the equation still holds because both sides have transformed in the same way. The equation is also form invariant with respect to recalibrations of length, although each side is not quantity invariant in respect of this transformation. Both sides *are* quantity invariant for displacements of origins and for linear boosts, so the equation must be form invariant for these. Finally, if we consider recalibration of time, we find that the left-hand side is not invariant but the right-hand side is. The equation is thus not form invariant under this transformation.

We have seen how the principle of relativity can be expressed in terms of form invariance (Section 4.1 of Unit 3). All laws of physics are form invariant under all transformations between equally calibrated inertial frames. The equation used in the above example passes this test. (The fact that the equation is not form invariant under time recalibration is immaterial as recalibrations of time and length are specifically excluded in the principle of relativity.) Of course, the mere fact of the equation passing the test by no means guarantees that it is a law, but any equation failing the test on just one transformation is a non-starter.

Finally, suppose that a well-established physical law is found to fail the form-invariance test. Should we conclude that it is not really a law? Perhaps, but first we should critically examine how we have transformed the quantities. The Lorentz force law in the form

$$\frac{\mathrm{d}\mathbf{p}}{\mathrm{d}t} = q(\mathbf{E} + \mathbf{v} \times \mathbf{B})$$

is a prime example, being firmly established experimentally but apparently not form invariant under linear boosts. But if the *relativistic* transformations are performed ($\mathbf{p}$ to $\mathbf{p}'$, $\mathbf{v}$ to $\mathbf{v}'$, t to t', $\mathbf{E}$ to $\mathbf{E}'$ and $\mathbf{B}$ to $\mathbf{B}'$) then it turns out that form invariance is regained.

9.3   Appendix 3: Application of the principle of relativity

Questions in Part II of the examination may ask you to apply the principle of relativity in one of two ways: to investigate the forms of physical laws or to make predictions about experimental observations. Here are two questions, with solutions, illustrating such applications. Such questions are likely to form part of a longer question. Note that in both these questions you are *not* required to use any *particular* physical laws apart from those given in the question. You might be able to obtain the result required in some cases by using, say, Newton's second law, but you would not obtain full credit unless you apply the principle of relativity correctly. It is a case of too much knowledge being a possible impediment! In the solutions, comments in square brackets are explanatory and not essential for a good answer.

Exercise A   An observer at the origin of an inertial frame S studies the motion of two particles which interact with each other but not with the rest of the Universe. He claims to discover the fundamental law

$$\mathbf{a}_1 = (A\mathbf{v}_2 - B\mathbf{v}_1)t^{2-\gamma}/|\mathbf{x}_2 - \mathbf{x}_1|^3, \tag{3}$$

where $\mathbf{x}_1$, $\mathbf{v}_1$, $\mathbf{a}_1$ are the position, velocity and acceleration of particle 1, and $\mathbf{x}_2$, $\mathbf{v}_2$ are the position and velocity of particle 2. A, B and γ are universal constants whose values do not depend on the time t, or on the way the particles are set in motion.

Assuming that Equation 3 is form invariant under any linear boosts and translations in time, write down the corresponding law in a second inertial frame S', whose spatial origin coincides with that of S at $t = 0$ but whose time origin is shifted by an arbitrary amount τ, and which is moving at an arbitrary but constant velocity $\mathbf{u}$ with respect to S. Hence, use the principle of relativity to show that the proposed law is inconsistent with Newtonian ideas of space and time *unless* $A = B$ and $\gamma = 2$. Show also that the proposed law is *not* in general form invariant under recalibrations, and briefly comment on the application of the principle of relativity in respect of this absence of invariance.

Solution   A law is form invariant under a certain transformation if it can be written with the same arrangement of symbols in any frames related by the transformation. So in S′ the law would be written:

$$\mathbf{a}'_1 = (A\mathbf{v}'_2 - B\mathbf{v}'_1)t'^{2-\gamma}/|\mathbf{x}'_2 - \mathbf{x}'_1|^3. \tag{4}$$

[We have not applied a transformation yet. We cannot actually do this without making assumptions about time and space, which we do in the next part of the solution.]

According to the principle of relativity, all physical laws in inertial frames *are* form invariant with respect to linear boosts and time origin shifts. Suppose space and time are Newtonian, then $\mathbf{x}'_1 = \mathbf{x}_1 - \mathbf{u}t$, $\mathbf{x}'_2 = \mathbf{x}_2 - \mathbf{u}t$, and $t' = t - \tau$. Since $\mathbf{v} = d\mathbf{x}/dt$ it follows that $\mathbf{v}'_1 = d\mathbf{x}'_1/dt' = d(\mathbf{x}_1 - \mathbf{u}t)/dt$ since $dt/dt' = 1$, τ being constant. Thus $\mathbf{v}'_1 = \mathbf{v}_1 - \mathbf{u}$ and $\mathbf{v}'_2 = \mathbf{v}_2 - \mathbf{u}$. Similarly, since $\mathbf{u}$ is constant, it follows that $\mathbf{a}'_1 = d\mathbf{v}'_1/dt' = d\mathbf{v}_1/dt = \mathbf{a}_1$.

(You would get most of the credit in this part if you simply stated that acceleration is an invariant quantity under linear boosts and time translations, and that velocity transforms by the Galilean equation $\mathbf{v}' = \mathbf{v} - \mathbf{u}$, both assuming Newtonian space and time.)

Substituting into Equation 4 (you could equally well substitute the inverse transformations into Equation 3) gives

$$\mathbf{a}_1 = (A\mathbf{v}_2 - B\mathbf{v}_1 - (A - B)\mathbf{u})(t - \tau)^{2-\gamma}/|\mathbf{x}_2 - \mathbf{x}_1|^3, \tag{5}$$

which only agrees with Equation 3 if $A = B$ and $2 - \gamma = 0$, i.e. $\gamma = 2$. Hence these conditions are needed for the proposed law to be consistent with Newtonian space and time.

Recalibrations are achieved by the transformations $\mathbf{x}' = \lambda\mathbf{x}$, $t' = \mu t$. Using definitions of velocity and acceleration we find that $\mathbf{a}' = \lambda\mathbf{a}/\mu^2$ and $\mathbf{v}' = \lambda\mathbf{v}/\mu$. Substituting into Equation 4 gives

$$\lambda\mathbf{a}_1/\mu^2 = (\lambda/\mu)(A\mathbf{v}_2 - B\mathbf{v}_1)(\mu t)^{2-\gamma}/(\lambda|\mathbf{x}_2 - \mathbf{x}_1|)^3$$

or

$$\mathbf{a}_1 = (\mu^{3-\gamma}/\lambda^3)(A\mathbf{v}_2 - B\mathbf{v}_1)t^{2-\gamma}/|\mathbf{x}_2 - \mathbf{x}_1|^3$$

so that unless $\mu^{3-\gamma} = \lambda^3$ the constants A and B will be different in a recalibrated frame. The proposed law is thus not form invariant under general recalibrations. The principle of relativity specifically excludes recalibrations from the list of inertial transformations under which laws must be form invariant, thus no further aspects of the proposed law can be tested in this way.

Exercise B An isolated physical system, consisting of three particles A, B, and C, is viewed from an inertial frame S. Particles A and B are identical and have mass m. Particle C has mass $m/3$. The initial positions and velocities are

$$\mathbf{x}_A(0) = (3, 4, 8) \qquad \mathbf{v}_A(0) = (2, 0, -1)$$
$$\mathbf{x}_B(0) = (3, 4, -8) \qquad \mathbf{v}_B(0) = (2, 0, 1)$$
$$\mathbf{x}_C(0) = (0, -2, 0) \qquad \mathbf{v}_C(0) = (3, 1, 0).$$

Write down the initial positions and velocities in a frame S′ obtained from S by a reflection of the **3**-axis, and use the principle of relativity to explain why particle C *cannot* leave the plane $x^3 = 0$.

Solution   To transform by a reflection of the **3**-axis, simply reverse the **3**-components of all quantities:

$$\mathbf{x}'_A(0) = (3, 4, -8) \qquad \mathbf{v}'_A(0) = (2, 0, 1)$$
$$\mathbf{x}'_B(0) = (3, 4, 8) \qquad \mathbf{v}'_B(0) = (2, 0, -1)$$
$$\mathbf{x}'_C(0) = (0, -2, 0) \qquad \mathbf{v}'_C(0) = (3, 1, 0).$$

Because A and B are identical particles, the descriptions of the system in S' and S are identical. Since the system is isolated and the frames S and S' are both inertial and equally calibrated, the principle of relativity predicts that the subsequent descriptions of the system in the two frames will continue to be identical. Now if particle C were to acquire say a positive x^3-component in S, the corresponding component x'^3 in S' would be negative so the descriptions would not then be identical. So particle C must stay confined to the plane $x^3 = 0$.

[The principle of relativity as originally stated in Unit 3 related to different experimenters observing *different* isolated systems. However, in this case they actually observe the *same* system and their identical descriptions are due to the symmetry of the system. An explanation along the lines of 'symmetry begets symmetry' would earn some credit. However, if for example you had invoked forces — you might have correctly argued that the **3**-components of the total force on C due to A and B cancel — then this explanation, not being in the spirit of the question, would earn little credit.]

9.4   Appendix 4: Invariance and conservation

That these two terms are sometimes confused is due entirely to the similarity of their non-technical meanings. The technical meanings, in the context of this course, are absolutely distinct, so if you have a tendency to use these terms interchangeably, read on.

Invariance, whether of a quantity or an equation (form invariance), is always associated with *transformations between frames of reference*. Thus, for example, a quantity, or a law, may be invariant under origin shifts, or linear boosts. By contrast, conservation is associated only with quantities which *do not change over any period of time*, as measured in a particular (usually inertial) frame of reference. Conservation always involves the concept of an *isolated system* which may have internal interactions but no input from outside of energy, charge and so on. Given a particular transformation, the invariance of any quantity or equation can be investigated mathematically, without reference to any laws, as was explained in Appendix 2. Conservation is investigated experimentally and enshrined in fundamental laws.

The following examples, which all relate to Newtonian spacetime, may help you to consolidate your appreciation of the distinction:

The distance between two particles is invariant under linear boosts, but is not conserved.

The linear momentum of a system is not invariant under linear boosts, but is conserved.

The kinetic energy of a system is neither invariant under linear boosts nor conserved.

The mass of a system is both invariant under linear boosts and conserved.

(Which *one* of these statements does not survive the transition to the spacetime of special relativity?)

9.5   Appendix 5: Mathematical hazards

All parts of S357 require some proficiency in mathematics, but Block 3 is the most demanding. These notes are written mostly with Block 3 in mind but they should be helpful for the whole course.

Brackets (parentheses)

Page and equation references here are to Unit 10 & 11.

These are used liberally, in order to make equations quite explicit, but unfortunately this makes many equations longer than strictly necessary. Brackets are used in three ways, all of which occur in Unit 10 & 11. The first way, exemplified by the equations at the foot of p. 8, is as a means of *specifying a vector quantity* by grouping the components together. This is unlikely to cause confusion as the symbols within these brackets are separated by commas. Moreover, these brackets are rarely used with either of the other types in the same equation, or at least on the same side of an equation. However, the other two uses occur together very frequently, and then confusion can arise. Equations 3a and 3b (p. 19) both use only *associative brackets*. These indicate that operations within a pair of brackets must be carried out first, before any outside operations. So the subtractions within the round brackets must be done before the square is found, and the square brackets signify that all the squared terms must be added before the square root is found. The third use of brackets is to *show functional dependencies*. All brackets in the equations on p. 17 before SAQ 2 are of this type. They simply mean that each component of $\mathbf{x}$ is a function of time or of angle as the case may be. The functional brackets and the symbol(s) contained within (the 'arguments') may be omitted without invalidating the equation. (A quantity that is a function of two or more arguments would be indicated by functional brackets containing the symbols of each variable, separated by commas, e.g. the left-hand side of Equation 11; indeed P and Q would be better described as 'labels' in this case.)

Now look at the first equation on p. 10. Here vectorial brackets, the outer brackets on the right-hand side, appear together with functional brackets enclosing t_0. The functional brackets indicate that each component, regarded as a function of time, is to be evaluated at the instant t_0, and the resulting vector is the position vector for the same instant. In Equation 8b (p. 22) you will find functional and associative brackets in close conjunction. The functional brackets and their contents can be omitted and this has been done in Equation 8c.

Which are the functional brackets in Equation 33 (p. 49)?

The brackets immediately following g_2 and enclosing q^1. The others are all associative and indicate what is to be squared or 'square rooted'. Of course $g_2(q^1)$ *could* be read as g_2 times q^1, i.e. the functional bracket misread as associative. That this is *not* the case should be clear from the context, but the mistake is understandable if you look at the equation in isolation. The following margin note explains that the notation has been simplified in Equations 34 and 35. If you have trouble distinguishing between functional and associative brackets, it would be worth rewriting all relevant equations with the functional brackets (and their contents) removed.

Once you are confident about the purposes of the brackets, you should have no difficulty in identifying metric coefficients, etc. in any given expression, when you have been provided with the standard formulae. The mathematical tasks you may be given are rather limited — you might be asked to find the curvature function for a given metric, or to check whether (or not) a certain line in the surface (expressed in terms of coordinate functions) satisfies the geodesic equations. You will *not* be required to solve the geodesic equations for any non-trivial metric since this could be a major mathematical task.

The following tips may help you to recognize and handle the commonest hazards:

1 In differentiation (or integration) be very sure which quantity you are differentiating (or integrating) with respect to, which quantities are functions of this, and which quantities are constant. You will not have to deal with exotic or complicated functions. Integration blunders can be spotted if you differentiate your answer — you should of course regain the original function.

2 Do each step on a separate line, as a separate equation. For example, write:

$$\sin \theta = \tfrac{1}{2}$$
$$\theta = \sin^{-1} \tfrac{1}{2}$$
$$\theta = \pi/6$$

rather than $\sin \theta = \tfrac{1}{2} = \pi/6$.

3 Look for simplifications such as cancellations, but avoid short cuts. If you are asked for a numerical answer, delay substituting numbers until the last possible moment. Cancellations are much more obvious with symbols than with numbers; moreover, this strategy minimizes the number of keys you need to press on your calculator!

4 Leaving an answer in the form of a square root (for example, $\pi/\sqrt{3}$) is normally acceptable and saves time and the possibility of calculation error.

5 When you arrive at a numerical answer, take a moment or two to consider whether or not it is reasonable. If it is, this does not of course mean that it is correct. But a completely unreasonable answer obviously indicates a blunder somewhere. For instance, if you were asked to find the distance to a particular cluster of galaxies, and your value is $6 \times 10^{24}\,\text{m}$ you might not know if this was right, but if your value is $6 \times 10^{-24}\,\text{m}$ then you should certainly know that it is wrong! If this sort of thing happens then check that you have manipulated powers of ten correctly, and that you haven't multiplied when you should have divided (or vice versa). Often a rapid check can reveal such a blunder. If not, and time is short, then add a suitable comment to leave the examiners in no doubt that you have grave reservations about your final value.

6 The right-hand side of an equation is often a square or square root: don't overlook this.

7 If you can, check that the dimensions of your answers are consistent. The dimensions of curvature are $(\text{length})^{-2}$. Coordinates q and metric coefficients g could have various dimensions depending on the coordinate system used, but gq^2 is always the square of a length. In the Robertson–Walker metric, the scale factor $R(t)$ has dimensions of length, while the comoving coordinate σ and the spatial curvature parameter k are both dimensionless.

8 Where mathematical work involves vectors, full credit can generally be obtained only if you show awareness of this by correctly using vector notation. The bold typeface of the Units (which you may have emulated if your TMAs were word-processed) is not easily reproduced in the examination, so instead of **x** you should write $\underline{x}$. Other notations such as $\tilde{x}$ or $\vec{x}$ are equally acceptable. The point is that symbols representing vector quantities must be distinguishable from those representing scalars.

Of course, your manipulations should also be correctly notated. Products of vectors must be clearly shown as either scalar (dot) or vector (cross) products, and division must always be by a scalar, so use the modulus notation where needed (e.g. divide by $|\underline{x}|$ not by $\underline{x}$).

Don't be paranoid about this; in an exam you will always have higher priorities than hunting through your work checking that every vector symbol is underlined. The occasional slip will not lose you credit, but showing complete disregard for the vector character of your calculation probably will. In particular, examiners don't like equations where a vector is set equal to a scalar.

9.6   Appendix 6: General relativity and its allegory

Questions requiring descriptive answers based on Block 3 material may call for discussion of aspects of general relativity or of the two-dimensional 'allegory', or indeed of both. As it expounds the key ideas, Unit 10 & 11 sways from one to the other, and your answer may need to do something similar. The purpose of this Appendix is to alert you to the need to use *precise language* appropriate to the case you are discussing. Some terms are common to both, but many are not. If, for example, you use a term in discussing general relativity that should only be used in the allegory, you could well lead the examiners to suspect that you have not grasped the distinction. What follows should help you to avoid this situation.

Two-dimensional allegory	*General relativity*
surface (2 dimensions)	spacetime (4 dimensions)
point (x^1, x^2)	event (x^0, x^1, x^2, x^3)
line or curve	world-line
length	proper time τ
minimum length	maximum proper time
metric of the surface (4 metric coefficients in general but normally just 2)	metric of spacetime (10 metric coefficients in general but normally just 4)
curvature$_4$	Riemann curvature (20 components)
flat/curved surface	flat/curved spacetime
Euclidean plane	spacetime of special relativity
geodesic curve	geodesic world-line

Terms common to both

path
geodesic equations
geometric properties
local region
coordinate functions
parameters

9.7 Appendix 7: Working out a revision strategy

Have you worked out a revision strategy for yourself? You need one, because an aimless wander back over the course material will not be an efficient use of your time. The elements of your strategy should include, for each Unit, reading the Objectives and asking yourself which you can do, making sure you can answer the SAQs (or at least follow the solutions) and studying the Summary. Use all the Units in the course, including this one, as selectively as you need to. Reread the comments your tutor has written on your TMAs to remind yourself of your strengths and weaknesses and help you decide where to concentrate your effort. You know that the assessment is balanced fairly evenly between mathematical and descriptive work. If your strength is in the maths, concentrate on improving your descriptive work, and vice versa. It is always tempting to do more work in areas you enjoy and know you are good at, but resist the temptation because the 'law of diminishing returns' operates in a big way here. If you can get to a tutorial, go with a list of questions written out. If not, remember most tutors are happy to deal with questions by phone, e-mail or (preferably for mathematical questions) by post. Finally, do note that the exercises in this Unit do *not* represent a complete coverage of what is important. It is the Objectives and Summaries in each Unit and the continuous assessment that will guide you to a judgement of what you should know.

10 Solutions to consolidation exercises

Remember that all references by number to equations etc. are to the Units to which the Exercise is related.

10.1 Exercises for Units 9–12

Exercise 1 The answer is 'yes' for the question (and 'no' for the hint).

Any body released from rest in the station will, with the usual provisos about negligible effects of air in the space station, etc., move in a free-fall orbit around the central mass. Orbits of different radius will have different periods (remember Kepler's third law from Block 1), and so the objects will eventually move apart at a rate which depends on how different their distances are from the central mass. If the space station were travelling uniformly in outer space, anything released at rest within the space station would simply move along with it, so two objects released at rest in such a space station would remain forever at the same points in the space station and not move apart from each other.

Exercise 2 (a) The parachutist is not in free fall because she is subject to hydrodynamic forces due to the air ('wind resistance' in other words). (b) One member of the binary star is not in free fall because its mass is sufficient to make the other star recoil (they orbit around their common centre of mass). It cannot therefore be considered to be a 'test-body'. (c) The planet Mercury, however, *is* in free fall (...at least in the good approximations (i) that it is not significantly retarded by interplanetary material, and (ii) that the recoil of the much more massive Sun can be ignored).

Exercise 3 They will accelerate apart since bodies at different distances from the centre of the Earth will not have exactly the same accelerations.

Exercise 4 The equivalence principle tells us that what will be found in the space laboratory will be indistinguishable from what will be found in an effectively uniform gravitational field (i.e. at a suitable place near a concentration of matter). Since the laser beam is fired in the direction of the acceleration (from the 'rear' to the 'front' of the vehicle), this will be equivalent to firing the laser from the 'bottom' to the 'top' of a stationary laboratory on Earth (an occupant will not distinguish a force towards the 'rear' from a force towards the 'bottom'). Since we know from Unit 9 that light travelling upward in a uniform gravitational field will be received redshifted, so the light received at the 'front' of the accelerating laboratory will also be received redshifted.

Exercise 5 We use the equivalence principle by removing the Earth and giving the rocket an upwards acceleration of magnitude g. The light takes a time w/c to cross the laboratory, and in this time the laboratory travels upwards by $g(w/c)^2/2$ with respect to the inertial frame in which the laser was at rest when it emitted the light. Since the light will continue horizontally in the inertial frame, it follows that $P_1P_2 = g(w/c)^2/2$. Substituting values with $w = 100$ m gives a deflection of about 0.5×10^{-12} m, smaller than an atom but bigger than a nucleus. This must be the limit of accuracy since there is no conceivable test of straightness better than a light ray. Any ruler, however stiff, and any thread, however taut, will sag by a lot more than this when held horizontally.

Exercise 6 (a) A metric is an expression for the length of a short segment in terms of the small changes in the coordinates between the points at its ends. The Δ indicates the small changes, therefore the coordinates q^1 and q^2 are h and ϕ. In the simplified metric form we mostly use, q^1 is the coordinate with the constant metric coefficient, so it must be h with the metric coefficient $g_1 = 1 + k^2$. Thus q^2 is ϕ and the metric coefficient g_2 is h^2k^2.

(b) To evaluate Equation 68b, we need to find dg_2/dq^1, i.e. $d(h^2k^2)/dh = 2hk^2$. We also need d^2g_2/dq^{12}, i.e. $d(2hk^2)/dh = 2k^2$. Substituting these into the equation, and using the above expressions for g_1 and g_2, we find that

$$\mathscr{K} = \frac{1}{h^2(1 + k^2)} - \frac{1}{h^2(1 + k^2)} = 0$$

which is reasonable since we know that the curvature$_4$ of the surface of a cone is zero.

(c) To evaluate Equation 38c, we need $\dot{q}^1$ and $\dot{q}^2$. Firstly, $\dot{q}^1 = dq^1/ds = dh/ds$. Now a and k are constant but we must find $d(1/\cos s)/ds$ which is $\sin s/\cos^2 s$; hence

$$\frac{dh}{ds} = \frac{a \sin s}{(1 + k^2)^{1/2}\cos^2 s}.$$

Also

$$\dot{q}^2 = \frac{d\phi}{ds} = \frac{(1 + k^2)^{1/2}}{k}.$$

We also need g_1 and g_2 as before, and since g_2 involves h we must use the parametric expression for h in terms of s. We get

$$L^2 = (1 + k^2)\left[\frac{a \sin s}{(1 + k^2)^{1/2}\cos^2 s}\right]^2$$
$$+ \left[\frac{a}{(1 + k^2)^{1/2}\cos s}\right]^2 k^2 \left[\frac{(1 + k^2)^{1/2}}{k}\right]^2$$

which reduces to $a^2/\cos^4 s$ after cancellations and trigonometric identities have been used, leaving $L = a/\cos^2 s$.

(d) For Equation 38a, we need $g_1 \dot{q}^1/L$ which is $(1 + k^2)^{1/2} \sin s$. The first term in the equation is therefore $(1 + k^2)^{1/2} \cos s$. The second term, on substitution of expressions already obtained, will be found to cancel this, so the first geodesic equation is satisfied. In Equation 38b, $\dot{q}^2$ is constant, L is inversely proportional to $\cos^2 s$ and so is g_2 (because h^2 has this dependence). So the expression to be differentiated is independent of the parameter s and the second geodesic equation is also satisfied.

Exercise 7   According to the formula $C/r = 2\pi \sin \theta/\theta$, when $\theta \ (= r/R)$ exceeds $\pi/2$, C/r will start to decrease. When $\theta = 5\pi/6$, $\sin \theta = 0.5$ and $C/r = 1.2$ and as θ approaches π the circumference approaches zero. This circumference is bounding an area which approaches the total surface area!

Exercise 8   A narrow zone lying between θ and $\theta + \Delta\theta$ has a length $2\pi R \sin \theta$ and a width $R \, \Delta\theta$. Its area is thus $2\pi R^2 \sin \theta \, \Delta\theta$. The total area enclosed by the circle of constant θ is found by integrating this with respect to θ, starting from $\theta = 0$, giving $2\pi R^2(1 - \cos \theta)$. Since $R = r/\theta$, the required ratio is $2\pi(1 - \cos \theta)/\theta^2$. For very small angles, $\cos \theta$ is approximately $1 - \theta^2/2$, giving a ratio of π. This is expected, since the geometry of very small figures will approximate the geometry of a plane.

Exercise 9   This triangle covers the whole sphere, but for solving the problem we imagine a very small uncovered triangle (see Figure 2 of this Unit). Since this uncovered triangle is so small, the sum of its internal angles is π. Now, the internal angles of the covered triangle are the external angles of the uncovered triangle. The total angle round all three corners is $3 \times 2\pi = 6\pi$, so the required sum is $6\pi - \pi = 5\pi$.

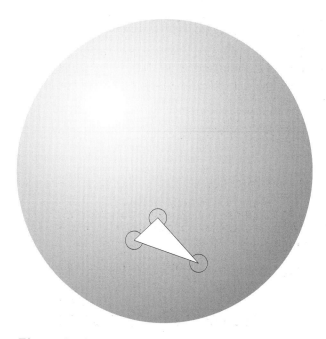

Figure 2   A *very large* triangle drawn on a sphere. The angles shown are the *interior* angles of the triangle.

Exercise 10   The Equator is the only line of latitude which is a great circle or geodesic. Hence the 'rectangle' must have at least one side which is not the shortest curve between corners, i.e. hence not a 'line'.

Exercise 11   Nothing can be at rest within the event horizon of a black hole — everything moves towards the central singularity. So the base of the tower would collapse, allowing the part above the event horizon to fall into the black hole also.

Exercise 12   A very distant observer, at rest with respect to the black hole, measures coordinate time t. Such an observer could see an event close to (but above) the event horizon, and would ascribe a time t to it (and a position r) according to the procedure described in Unit 6, Section 2.1 SAQ 4, i.e. assuming both the event and the observer are in the same flat inertial frame.

Unit 12 also discusses the proper time τ recorded by two other observers, both close enough to the black hole to be influenced by it, but outside the event horizon. For an observer at rest at constant radial coordinate r, $\Delta\tau = \sqrt{1 - k/r} \, \Delta t$ or $\Delta t = \Delta\tau/\sqrt{1 - k/r}$ (p. 6) which represents a large gravitational redshift seen by the distant observer. For a freely falling observer, $\Delta t \propto \Delta\tau/(1 - k/r)$ (Equation 8) which represents an even larger redshift — as would be expected for light from an object not only close to a centre of attraction but also retreating — and shows that the distant observer never sees the object reach the event horizon since Δt tends to infinity as r approaches k.

The fourth time measure used in Unit 12 is Eddington–Finkelstein time t' which can be used to rewrite the Schwarzschild metric so as to remove the apparent singularity at the event horizon (while retaining the central singularity). This reflects the fact that an observer falling through the event horizon would experience nothing special at that point. When plotted using Eddington–Finkelstein time, light cones, whose axis is the time axis at great distances, tilt progressively as the event horizon is approached and crossed in such a way that outward motion becomes impossible within the event horizon.

Exercise 13   (a) A star of radius say $1.2k$ would have, according to Birkhoff's theorem, exactly the same spacetime properties outside as any other spherically symmetric object of the same mass including a photon-sphere. It would not, however, have an event horizon because its radius is greater than the Schwarzschild radius k. Light transmitted radially upwards through the photon-sphere escapes, so the star could emit light in all directions. (The lines in Figure 8 are reversible until they cross the event horizon.)

(b) Reflection can also occur as follows: incident light with impact parameter less than $2.6k$ will pass through the photon-sphere and strike the surface of the star. Assuming that some of it is reflected at a variety of angles, those rays which are close enough to the outward radial direction will escape. Close enough means very close if the star's radius is only just over k, but up to nearly 90° as the radius approaches $1.5k$.

Whether seen by its own or by reflected light, the star would appear as a disk of radius $2.6k$.

The surface of a star of radius $1.5k$ would be a very strange place. Its two-dimensional geometry would be exactly like that of an ordinary spherical surface. However, it would *look* flat because light transmitted horizontally, i.e. tangentially, would remain horizontal — there would be no horizon. But because the light really orbits, you would see yourself in *every direction* at distances of $2\pi \times 1.5k$, $4\pi \times 1.5k$, etc! You could only see images of *another* object on the surface in two directions. Looking up, you would be able to see *every* visible star — Figure 8 shows that light can reach the photon-sphere from behind.

Exercise 14   The radius of the photon-sphere is $1.5k$ where k is the Schwarzschild radius of the black hole (Section 2.5). Light travelling radially *inwards* from infinity is *blue*shifted by the factor $(1 - 2MG/c^2r)^{1/2} = (1 - k/r)^{1/2}$. This can be seen by inverting the argument of Unit 10 & 11, Section 13.1, i.e. Δt is now the period of the emitted radiation and $\Delta\tau$ the period of the received radiation. For $r = 1.5k$, we get $\Delta\tau/\Delta t = \lambda(\text{received})/\lambda(\text{emitted}) = (1 - 2/3)^{1/2} = 1/\sqrt{3}$ giving $\lambda(\text{received}) = 600 \times 10^{-9}\,\text{m}/\sqrt{3} = 346 \times 10^{-9}\,\text{m}$.

The mass of the black hole is obtained from $k = 2MG/c^2$ giving $M = 4.5 \times 10^{34}\,\text{kg}$.

Exercise 15   The tidal force at the event horizon of a Schwarzschild black hole is proportional to the inverse square of its mass, see Unit 12, p. 24. It is imperceptible at the event horizon of a galactic-centre black hole but enormous at the event horizon of a stellar black hole. Any spaceship or astronaut would be torn to tiny pieces well before crossing this surface. For mini-black holes it is stronger still and can tear apart the particle–antiparticle pairs which form spontaneously but briefly from quantum fluctuations of a vacuum. This enables such a black hole to radiate (mostly photons and neutrinos) from *just outside* its event horizon, sufficiently to lose mass rather than gain, as stellar and larger black holes must do.

10.2 Exercises for Units 13–15

Exercise 16   Unit 13 describes five methods:

Cepheid variables: this is only applicable to galaxies close enough for these stars to be distinguished and their periods determined (up to $\sim 5 \times 10^7$ light-years).

Type Ia supernovae: being brighter than Cepheids, these events can be used to determine distances in much more distant galaxies, but they are comparatively rare.

Type II supernovae: rarer still than Type Ia supernovae but even brighter.

Radio interferometry: applicable only to certain radio galaxies.

Statistical method: applicable at any distance but only to galaxies in largish clusters where it can be assumed that the brightest galaxy has a particular absolute luminosity.

Exercise 17   From Equations 4 and 6 we have $z = Hr/c$, and putting $r = 5 \times 10^8\,\text{ly} = 4.73 \times 10^{24}\,\text{m}$ gives $z = 0.036$. Now $1 + z = \lambda_0/\lambda_1$ so that λ_1, the emitted wavelength, is $504\,\text{nm}/1.036 = 486\,\text{nm}$.

Exercise 18   No. $z \approx v/c$ would limit redshifts to less than unity, unless galaxies can recede faster than c. Consequently Hubble's law in the form of Equation 6 is similarly restricted.

We can rewrite Equation 4 more accurately as follows:

$$z = (\lambda_0 - \lambda_1)/\lambda_1 = \lambda_0/\lambda_1 - 1 = f_1/f_0 - 1$$
$$= \sqrt{(1 + v/c)/(1 - v/c)} - 1$$

which allows the redshift to grow indefinitely as v approaches c. This does not invalidate Equation 3 but clearly makes the left-hand side of Equation 6 more complex.

With a redshift of 2, $(1 + v/c)/(1 - v/c) = (1 + 2)^2 = 9$ which gives $v = 0.8c$. This approach assumes both the astronomer and the galaxy share an inertial frame.

As you saw in Unit 14, we prefer to describe the spacetime of the Universe as curved, with cosmological redshift caused by expansion of space rather than by the Doppler effect. Indeed galaxies, or rather clusters of galaxies, are all regarded as locally at rest. In Unit 14, an exact version of Hubble's law is given involving proper distance. Velocity, expressed as a rate of change of proper distance, can exceed c even though each galaxy is at rest in its local inertial frame.

Exercise 19   The present frequency of the microwave background peaks at around $2 \times 10^{11}\,\text{Hz}$ which corresponds to a wavelength $\lambda = c/f = 1.5\,\text{mm}$. At decoupling, the temperature of both matter and radiation was about 3000 K so the redshift since then has been about 1000. The wavelength of maximum intensity at decoupling was therefore about $1.5 \times 10^{-6}\,\text{m}$. This is in the infrared, but 3000 K is about the temperature of a tungsten filament and there would have been substantial high-frequency visible thermal radiation.

Exercise 20   Nuclear properties measured in laboratory experiments:

(a) Mass of proton and neutron (giving initial neutron to proton ratio).

(b) Probability of each reaction as a function of energy.

(c) Energy released by each reaction.

(d) Decay times for neutrons and tritons.

Astrophysical properties:

(a) Present average baryonic density of the Universe.

(b) Abundance ratio of deuterium, ^{4}He, ^{3}He, and ^{7}Li to hydrogen in interstellar gas.

(c) Mass, spectra and energy output of stars (so that stellar nuclear synthesis can be modelled under all known conditions).

Exercise 21 No, because light takes a finite time to travel to us. An image of a region of sky is not a snapshot of the distribution of sources at any instant. Since recorded science began, the only starlight to have completed its whole journey to Earth has come from some of the closest stars in our Galaxy. When we look at neighbouring galaxies we are seeing them at or before the dawn of our own species. The farthest galaxies are seen as they were before life began on our planet. The cosmic microwave background radiation tells us that the Universe was highly isotropic and homogeneous at the time of decoupling; since then structure has developed at a variety of scales but it is impossible to measure this over the whole visible Universe at any one time.

Exercise 22 The derivation of the $S^{-3/2}$ law (SAQ 2) assumes a static Universe. If the Universe were denser, but still flat and static, the number of radio galaxies exceeding a given received power would be multiplied by a constant factor, giving lines parallel to the displayed line, but displaced upwards according to the factor involved. In an expanding Universe, the light from more distant sources was emitted at times of greater density. The plot would thus descend from the higher lines to lower lines as the weaker, more distant sources are excluded from the sum, and the gradient would be steeper than indicated by the $S^{-3/2}$ law.

Exercise 23 The further away the cluster, the greater its recession rate, the ratio being given by the Hubble parameter. But over a long time period, any one cluster will itself become appreciably further away. So if the Hubble parameter is unchanged, the cluster must recede proportionately faster, which means that the clusters would not only be receding but also accelerating from one another.

If $R = At^n$ then $dR/dt = Ant^{n-1}$ and $H = n/t$. Then $dH/dt = -n/t^2$ giving $q = 1/n - 1$.

Exercise 24 In a critical universe $k = 0$, and for a matter-dominated universe $R(t) = At^{2/3}$. So the required metric is

$$(\Delta S)^2 = c^2(\Delta t)^2 - A^2 t^{4/3} \times$$
$$[(\Delta\sigma)^2 + \sigma^2(\Delta\theta)^2 + \sigma^2\sin^2\theta(\Delta\phi)^2].$$

World-lines of light pulses passing through the origin must be radial, so $\Delta\theta = \Delta\phi = 0$. For light pulses, $\Delta S = 0$.

For outgoing light, the metric becomes $c\,\Delta t = At^{2/3}\Delta\sigma$. At time t_1, $\sigma = 0$ and at time t_0, $\sigma = \sigma_0$, the coordinate of the distant recipient. After

rearranging the symbols, we can integrate between these limits:

$$c\int_{t_1}^{t_0} dt/(At^{2/3}) = \int_0^{\sigma_0} d\sigma$$

giving

$$A\sigma_0 = 3c[t^{1/3}]_{t_1}^{t_0} = 3c(t_0^{1/3} - t_1^{1/3}).$$

Now the proper distance of an event from the origin is $R(t)\sigma$ which, for the recipient at the time of receipt of the signal, is $At_0^{2/3}\sigma_0$. We use the expression for $A\sigma_0$ just obtained and find v by dividing the proper distance by the time interval $t_0 - t_1$, obtaining the required expression $v/c = 3(t_0 - t_0^{2/3}t_1^{1/3})/(t_0 - t_1)$.

For incoming light, σ decreases with time so that $c\,\Delta t = -At^{2/3}\Delta\sigma$. At time t_1, $\sigma = \sigma_1$ and at time t_0, $\sigma = 0$, the coordinate of the observer. The expression obtained for σ_1 is the same as that obtained before for σ_0. For incoming light, the relevant proper distance is that of the source at time t_1 which is $At_1^{2/3}\sigma_1$. As a result we get

$$v/c = 3(t_0^{1/3}t_1^{2/3} - t_1)/(t_0 - t_1).$$

This formula, on substituting $t_1 = 1$ and $t_0 = 2$ gives $v/c = 0.78$ for incoming light (which has to 'overcome' the expansion of the universe) while the first formula gives $v/c = 1.24$ for outgoing light, 'aided' by the expansion.

Exercise 25 (a) In a spatially flat universe, $k = 0$. The density in a radiation-dominated universe is $\propto R^{-4}$ (Equation 4), say $\rho = BR^{-4}$. So Equation 5 becomes
$(At^{-1/2}/2)^2 = (8\pi G/3c^2)B/R^2 = (8\pi G/3c^2)B/(A^2 t)$
which is satisfied if $A^4/B = 32\pi G/3c^2$.

(b) To find the pressure, we can substitute into Equation 1. The left-hand side is $-1/(4t^2)$. On the right-hand side, $4\pi G/3c^2$ can be written as $A^4/(8B)$ so the right-hand side becomes
$-A^4/(8R^4) - 3A^4 p/(8B) = -1/(8t^2) - 3p/(8\rho t^2)$
giving $\rho = 3p$ from which the result follows.

(c) q is the left-hand side of Equation 1 divided by $-H^2$. If $R = At^{1/2}$ then $dR/dt = At^{-1/2}/2$, $d^2R/dt^2 = -At^{-3/2}/4$ and $H = 1/2t$ so $q = 1$. More directly, simply substitute $R = At^{1/2}$ into $q = -R\ddot{R}/(\dot{R})^2$ where dot and double dot here represent first and second time derivatives.

(d) Since $H = \dot{R}/R = 1/2t$ then $t = 1/2H$. In such a universe, if t_0 is the present time, this is related to the present value of H by $t_0 = 1/2H_0$. Compare our matter-dominated universe in which $t_0 = \dfrac{2}{3H_0}$.

Exercise 26 The answer to SAQ 25, which involves an energy density of $10^{94}\,\mathrm{J\,m^{-3}}$, gives the time constant τ to be 1.3×10^{-34} s. The expansion factor in 10^{-32} s will be $e^{t/\tau}$ where $t/\tau = 10^{-32}/1.3 \times 10^{-34} = 77$. So the Universe will have expanded by $e^{77} = 2.6 \times 10^{33}$. [Of course, there is an element of speculation concerning these numbers.]

11   Notes for answers to essay exercises

Exercise 1

Introduction

What we have presented here are notes only; you would be expected to include these points in logically connected and readable prose. Feel free to use equations and graphs, and also any information gained from other sources.

Dynamics — study of motion of particles. Need to *describe* motion: space and time-scales, definitions of displacement, velocity. Need to *explain* motion (especially under gravity): force theories, geometrical theories.

Space and time

Time measured by clocks at rest. NM uses universal time. Simultaneous events in one inertial frame (IF) are simultaneous in all IFs. SR: not true for frames in relative motion. Consequences: moving clocks run slow (muon experiment), moving rulers shrink in direction of motion. Only Lorentz invariant S unchanged. But stationary rulers and clocks (in IF) measure space and time equally everywhere. GR: only locally true, because IFs — freely falling — are only local. In extended frame, stationary clocks run differently (gravitational redshift), rulers change length (loss of metrical significance).

Mass and energy

NM: mass additive, conserved, invariant. Total energy = potential energy (PE) + kinetic energy (KE) conserved (elastic collisions). Inelastic collisions: energy lost but mass unaltered. PE invariant but KE $(= \frac{1}{2}mv^2)$ varies with linear boosts. SR: mass invariant but not additive or conserved in inelastic collisions. Mass invariance like Lorentz invariance (energy/momentum : time/displacement). Momentum, magnitude $= \gamma m v$ rises with speed, stops particles reaching c, but energy $= \gamma m c^2$ continues to grow. Also $E = mc^2$ for particle at rest. Energy–momentum four-vector: $m^2 c^2$ is invariant; energy of system is cp^0 in frame where momentum is zero. GR: retained locally, but energy important source term generating curvature.

Matter and geometry

NM and SR: geometry Euclidean, matter has no effect, space and spacetime both flat. GR: matter has mass, i.e. energy, so source of spacetime curvature. Space and time coordinates lose metrical significance. Motion under gravity follows geodesics which are curved, gravitational force abolished. Spatial curvature very small in solar system, orbits result from larger spacetime curvature (four-dimensional geometry). Near black holes space and spacetime highly curved so dynamics strange, e.g. compulsory inward motion, compulsory rotation.

Cosmology

Cosmological principle (homogeneity, isotropy). Space and time given by Robertson–Walker metric, with parameters k and $R(t)$ satisfying Friedmann equations, derived from GR. Comoving coordinates: universal frame exists where all clocks keep same time (though moving apart). But spatial properties can change as Universe expands. Energy density a universal quantity at any time, determines evolution of Universe.

Conclusion

GR most comprehensive but impractical except for high-symmetry cases (Schwarzschild and Robertson–Walker metrics). Not yet combined with quantum theory so dynamics very close to black hole and big bang singularities unknown.

Exercise 2

Introduction

Historical perspective: central role of mechanics in development of modern science. Motion on Earth and in heavens, Copernicus, Kepler, Galileo, Newton. 19th century advances (electromagnetism, precision optics) show spacetime of NM incompatible with principle of relativity. Speed of light invariant. Einstein proposes SR — new spacetime with new mechanics. Invariance of c, laws of mechanics and electromagnetism Lorentz invariant. Then tackles gravity — coincidence of the equality of inertial and gravitational mass. Proposes GR but this has little impact for about 50 years.

(i) Relationship between theories

NM works for low speeds $\ll c$ and weak gravity $r \gg GM/c^2$. In astronomy: planetary systems, stars within galaxies, galactic haloes, clusters. On Earth — almost everything except high-energy particles but can't relate electromagnetism in different frames. SR extends to speeds approaching c (and equalling c for light), but gravity still weak. Good accuracy for everything on Earth and in 'local' astronomy and cosmology (e.g. neutrinos, cosmic rays), black holes excepted, but has to use Newtonian gravitation. GR generalizes SR by allowing curvature of spacetime caused by energy and gives more complete description of gravity, even when very strong. But computationally intractable except when symmetry is high. Essential for describing black holes (very strong gravity) and evolution of Universe (large length and time-scales make GR essential). SR has been developed to include quantum theory — needed for sub-atomic scales; GR has not.

(ii) Evidence for SR

Decay of muons in storage ring (and in descent from upper atmosphere) confirms 'moving clocks run slow'; whereas in NM frames in relative motion have same time-scale.

Limiting speed c of high-energy particles in accelerator shows $E \to \infty$ as $v \to c$. In NM, $E = \frac{1}{2}mv^2$ so must be wrong. Change of total mass of particles in nuclear reactions/decays, Michelson–Morley experiment, c independent of source speed in pion decay, binary stars.

(iii) Evidence for GR

Gravitational redshift measurement of Pound and Rebka shows rate of clocks affected by gravity which is denied by NM. Accurate experiment using Mössbauer effect. Good agreement with GR but too local to be rigorous test, but it does test EP. Redshifts from some white dwarf stars bigger but test less accurate.

Bending of light and radio waves skimming Sun. Delays in radar pulses skimming Sun. Spacetime shown to be curved. Results consistent with GR. Residual perihelion advance of Mercury; change in binary pulsar period (Hulse–Taylor) through gravitational radiation; gravitational lensing; Gravity Probe B.

Conclusion

All three theories useful and necessary. GR most complete but avoided except for cosmology, black holes, binary pulsars, and measurements of highest accuracy as NM or SR otherwise satisfactory.